EVOLVED

CHRONICLES OF A PLEISTOCENE MIND

by Maximilian Werner

First Torrey House Press Edition, June 2013

Published by Torrey House Press, LLC
P.O. Box 750196
Torrey, Utah 84775 U.S.A.
www.torreyhouse.com

International Standard Book Number: 978-1-937226-17-6
Library of Congress Control Number: 2012955571

Figure 2: Araneidae stabilimentum by Madison Shino
Cover design by Jeffrey Fuller, Shelfish • shelfish.weebly.com

For Wilder

CONTENTS

List of Illustrations

FOREWORD

Charles Darwin's great ideas have been slow to take hold outside the arena of academic biology. Indeed, Gallup polls from 1982 to 2012 show that upwards of 40% of United States respondents believe that God created people in essentially their modern form, while another 40% or so are willing to accept the proposition that humans have evolved—provided that God got us rolling down the evolutionary path. I am one of those in the minority and as an evolutionary biologist, I applaud Maximilian Werner for his willingness to apply evolutionary ideas (without recourse to divine intervention of any sort) to help explain why we (and other creatures) behave the way that we do. In this, he joins not only the great many biologists who study evolution but also a coterie of anthropologists, psychologists, economists and even some social scientists and doctors who have recognized the value of asking whether a particular attribute might have evolved by natural selection.

The beauty of this approach lies in the breadth of issues that are open to Darwinian analysis and the ability of scientists to put Darwinian ideas to the test. As Maximilian Werner illustrates in this remarkable memoir, a great variety of phenomena, even those of a personal nature, can be examined profitably from this perspective. How a black widow designs its web, the evaluative interest we humans take in the landscapes we enter, whether we sleep on our sides or our backs, how we play with one another as children, the way in which we bury our dead—all these matters and a host of others take on an intriguing freshness once we wonder if these characteristics have been shaped by natural selection. The awareness that it is not enough to say that we and the organisms around us simply like or want to do things in particular ways leads Werner to look for the possible connections between the immediate causes of our behavior and the evolutionary causes of our actions. In the past, differences in the genes, neurons, emotions, likes, and dislikes of individuals have surely had behavioral effects resulting in differences in reproductive success. If so, our genes, neurons, emotions, likes and dislikes have evolved, predisposing us to behave in ways that helped our ancestors reproduce more than others of our species.

One does not have to agree with every evolutionary speculation of the author to see how the approach opens up a new vista on the natural world, particularly with respect to our own behavior and psychology, which so often are considered to be the all but random products of cultural mores. What Werner has adopted from the academic students of evolution is the capacity to propose novel hypotheses about the natural world and even his own actions, as he seeks to link his feelings and those of other people to the behavior and psychology of all the other evolved organisms on our planet. The realization that we have a history shaped by the same processes that have produced the black widow spider and the warthog is a liberating concept, as Werner shows us with example after example in his graceful, thoughtful prose. Even those emotions that we may consider negative, like a child's desire to kill harmless birds, can, when scrutinized by an evolutionary thinker, provide insights that no other approach can offer.

Almost all biologists and many psychologists understand the importance of the fact that our brains evolved in the Pleistocene epoch under relentless selection for attributes that promoted the survival and reproductive success of our ancestors. Werner brings this understanding to his readers in a highly personal fashion as he uses his own history and emotions to make evolutionary analyses real and meaningful. Readers of this book are likely to change how they view themselves and the world around them. I hope that readers realize that the book's arguments also constitute a compelling case for caring for the environment, for a complete world is an essential ingredient if we are to understand evolution and truly know ourselves. So off you go to accompany Werner on a journey that will take you to his backyard, to the desert, to his family, to his own childhood, a journey full of enjoyment, insights, and literary pleasure, a journey well worth taking.

—John Alcock

ACKNOWLEDGMENTS

Although I physically wrote this book in a year and a half, I've been writing it mentally my entire life. As the person who introduced me to the world and has supported me ever since, my mother, Nancy Werner, deserves a heartfelt thanks for helping me to realize this project. I must have read the entire book to her over the phone. Her willingness to simply listen would have been enough, but she often provided crucial feedback that helped me to make the book more appealing to the general reader.

Writing this book required me to become a student again. The world itself is a great teacher, but it always helps to know people who can clarify and deepen one's understanding of those teachings. Thanks to Kirsten Allen and Anne Terashima for their careful readership and to John Alley for helping me portray a more accurate picture of Native Americans. I also need to thank my wife, Kim. Without her support and love for me as well as for Darwin, parts of this book would not have been written.

I am deeply grateful to John Alcock for his ability to communicate complex ideas beautifully, simply, and directly—a skill he no doubt heavily relied on when discussing biology with me. Whether those discussions took place while hiking in the Sonoran Desert, or in the margins of my manuscript, John was always patient, supportive, and interested. Bert Bender is one of the most extraordinary teachers and men I have known. Because he opened my eyes to Darwin and his astonishing theory of evolution, I have hope that humanity might someday know itself and do right by this amazing Earth.

EVOLVED

CHRONICLES OF A PLEISTOCENE MIND

by Maximilian Werner

It’s a precious thing to bear.

–Stone Gossard

WANDERING AND WONDERING: A NEW STORY FOR THE WORLD

We will not serve what we do not love.
And we cannot love what we do not know.

—Loyal Rue, *Everybody's Story: Wising Up to the Epic of Evolution*

Humanity is exalted not because we are so far above other living creatures, but because knowing them well elevates the very concept of life.

—E.O. Wilson, *Biophilia*

This is a love story.

Although I do indirectly refer to sexual, courtly, and platonic forms of love in these pages, I am mainly interested in telling the story of the complex attraction that includes yet reaches beyond our typical understandings of these feelings and our expressions of them. This inclusiveness is what distinguishes ecocentric or biotic love from anthrocentric or strictly human forms of love. Love in this deep sense describes humans, but it is also shared by and extended to other animals and to the environment. My wife Kim's mother-love for our sixteen-month-old son, Wilder, is powerful and familiar, but so too is the pleasure she feels when she watches the sun rise in the mountains or crouches on a grassy river bank—fly rod in hand—sight-casting to brown trout as they devour mayflies in the late spring light. Wilder does not yet know the word love, but when he waves "hello" to the moon, or follows a honey bee from flower to flower as it probes for nectar, or points excitedly at an ant carrying a lacewing like a panel of lead-paned glass, the expression on his face suggests he knows the feeling firsthand. Love and awe and wonder: they're all wrapped up in his tiny body. I share his loves, but I have others, too, like rain when I have shelter and fire when I am cold. I seek out rivers when I want to rejoin other life and remember my own. I eavesdrop on an alpine meadow; ponder a pod of killer whales hunting seals in the dark waters of the North Pacific; I savor the dark pink flesh of wild salmon when I am fortunate enough to eat it, and I am momentarily stunned by the ruby conflagration that is an Anna's Hummingbird's throat when the sun hits it just right.

We take pleasure in certain things partly because we know their opposites. The sight of the rising sun gives us pleasure because we know night. The sight of food gives us pleasure because we know hunger. The complex and deep-seated connections we have to the little piece of ground we call home should remind us that the world as a whole is a life-giving and loveable place, deserving of our most ardent interest and care. The connection springs from our primordial relationship with the physical world, and it is as much the province and provocation of science as it is the quandary of philosophers, the revelation of mystics, and the inspiration of poets and rock stars. It demands that we bring our entire being to bear on the question of what it means to be human, to live better and more wisely on what has become an increasingly imperiled planet. But to do this, we've got to do a better job of exploring and telling the truth about human nature and the nature of existence.

This is why our stories are so important. As the vehicles of our values, they teach us our place in this precious and difficult world. Humility, reverence, indifference, domination, and hostility are familiar themes in our individual and collective narratives. What is clear, however, is that regardless of the values we may communicate through our separate stories, we all will decide what kind of world we will inhabit, one that is characterized by its integrity, health, and diversity—including a humanity whose respect for the finite world is the guiding principle of our conduct—or a world of dust and impoverishment and horrific violence: life without birdsong, forests without a single bear bed or spider web or uncontaminated drop of dew.

I enjoyed many natural wonders in my childhood, and I know life is worth living as long as these wonders remain. When I contemplate the dominant stories of North America, however, it does not seem to me that we have an especially enlightened view of nature and of ourselves. The marketing slogan for the *X Files*—perhaps one of the most popular shows in the history of television, spawning a vast array of popular successors—is illustrative. It assured viewers that "the truth is out there." If all we were talking about were preternatural plots and alien abductions and conspiracies, I wouldn't think twice about the show's 25-million-member audience. But in the context of the real world, an audience of this magnitude is unsettling, if only because the "truth" to which the show refers is not here, but "out there," flying around in an alien spacecraft, or emblazoned in moon tableau, hopelessly locked in enigmatic celestial writ. In reality, the truth may be unknown to us at the moment, but it is not alien, nor is it a conspiracy: It is we and we are it and it is everything, a vast community of species spread across the planet in a web of amazing variety and tantalizing complexity.

Fortunately, our scientific interests are many, and the world—our gorgeous, mysterious, and fleeting subject matter—is unquestionably real and accessible. (Who needs Martians when we can ponder the natural history of a marbled diving beetle, or the courtship behavior of the satin bowerbird?) Often, however, we seem paralyzed by the world's grandeur and mystery. Perhaps it is enough that we simply have these experiences of awe and wonder. Who cares what—if anything—lies behind them? Why is it important that we know? Given our undeniable emergence from, connection to, and dependence on the environment, I think we are wise to ask these kinds of questions.

Having a robust sense of mystery is natural and necessary to realizing our potential, but when it distances us from the real challenges we face, it is debilitating and destructive. This seems especially true when it comes to unassailable mysteries, or those mysteries in which we readily believe but for which no evidence or data are available. Under these circumstances, enlightened solutions to our personal and collective struggles don't come easily, if at all. Our belief in subjective truth may be personally fulfilling as far as our own definitions are concerned, but the belief may isolate us from other dimensions of ourselves and the world we inhabit. I believe we are never more informed, fulfilled, and human than when we attend to our complex, wonderful, and sometimes troubling nature, particularly as it relates to the physical world.

How can we hope to understand such mind-boggling intricacy and complexity? What tool is best suited for the job? I share the long-standing, well-established, and widely accepted feeling that the epic of evolution is the truest means we have for addressing these questions. This is not a matter of pitting one story or world-view against another. It is a matter of rethinking our priorities in the context of what we know to be true about the planet and ourselves. If our morality includes a commitment to posterity, don't we have an obligation to preserve the planet's health and richness, the very conditions that make our lives and stories possible? If so, what on Earth is stopping us?

I explore these kinds of questions throughout the book, but let me offer two abbreviated and interrelated reasons for our apparent stubbornness. The first is that, like most other organisms on the planet, we tend to be biologically conservative. Whether it comes to our ideas and the language we use to express them, or our interests in music and food and mates, we usually stick with what we know—"the usual," as one recent fast food commercial put it—unless we are forced or think to do otherwise. Change is difficult for us, and beyond fulfilling our day-to-day needs, we guard our time and energy. That we are naturally conservative is also reflected by the relative simplicity of our major cultural and religious narratives, which appear to posit unassailable mystery as a condition of religious devotion and conviction.

Certainly, some degree of demystification occurs when we use science to understand the natural world and ourselves, but that does not make the subject—whatever it may be—any less amazing. Our increased understanding and knowledge of how things work not only makes the world more wonderful, it empowers us to make better decisions about how to live in it. Science does not have all the answers, of course, but instead of viewing the mysteries of existence as inscrutable, science views mystery as an impetus and invitation, in the same way that a puzzle is both mysterious and inviting when its pieces lie face down and scattered on the floor.

As someone who is trying to figure out my place in this astonishing world, the notion of ghosts passing through walls or aliens crashing in the desert or gods intervening on behalf of certain humans and not others would be amusing if it weren't so tragic. As I write, the world is in the throes of an unprecedented environmental, social, and religious crisis. Within the last two months, 8,000 people have been killed in sectarian violence in Syria. My journalist friend tells me that, to date, well over 100,000 Iraqi civilians, many of them children, have lost their lives since the United States invaded Iraq. Now the focus is the war in Afghanistan, a conflict with no end in sight.

Here in the U.S., PBS closes the news hour each night with

photos of our dead, sometimes ten to fifteen at a time, the majority of whom are smooth-faced boys in their late teens and early twenties. Regardless of who dies, whether soldier or civilian, war is a crime against humanity and is one of the greatest preventable failures of our species. This is to say nothing of the crimes we've committed, and continue to commit, against the nonhuman world in the name of progress. Writer Linda Hogan once commented on how our conflicts with the environment and with each other stem from the conflicts we have within ourselves. I think she's right. I know that our reasons for doing what we do are complex, but if this is where our current stories have gotten us, I think we need new stories about our place in the world.

Scientists and naturalists have long applied evolutionary theory to any number of different organisms in an effort to understand how they have come to live and behave in the ways they do. In recent years many of the hard, soft, and social sciences—including cellular biology, ecology, psychology, and sociology—have been reinvigorated by evolutionary theory. More importantly, perhaps, is that the humanities have been resuscitated (and, ultimately, grounded) by the publication of several works that show evolutionary theory's relevance to the arts, religion, and other arguably transcendental and distinguishing characteristics of our species. In general, scientists and humanists who do science know that their theories are valuable to the extent that they can be tested and thus explain and predict phenomena. This is no less true for those working in the evolutionary sciences.

In fact, Darwin's theory can be applied in any number of interesting, surprising, and worthwhile ways. The fact that my own approach is personal and speculative has not come without its risks and challenges. I draw on current research, but my goal is not to argue a single claim and marshal a body of evidence in support of it. Instead, I use personal narrative and memoir to tell the story of how Darwinian evolution might be used to improve and deepen our understanding of day-to-day life in all its forms.

Years before I started this project, John Steinbeck alerted me to the challenges of applying the all-encompassing theory of Darwinian evolution. In *Log from the Sea* of Cortez he cites the "looseness" of nonteleological thinking, and warns that one must avoid the various hazards associated with so much freedom and flexibility. To this end I have had some great teachers. Through my associations with Darwin scholar Bert Bender and biologist John Alcock, as well as through my own vigilance, I have made every effort to avoid distorting or confusing evolution's possible relevance to our lives.

We are descended from humans who emerged during the Pleistocene epoch, so what we are as a species is the direct result of this ancient lineage. If we wish to know more about our origins, we can begin by looking no further than the mirror. On the evolutionary timeline, then, each moment is the moment of the Pleistocene mind. I chronicle an expanse of time ranging from my age at the time of writing to roughly 2.7 million years ago, which is around the time the Pleistocene epoch began. Rather than starting at some point in the recent past and moving forward in time, I begin in the present and work my way back through deep history, which is where speculation proves very handy. Although I do not often draw definitive conclusions, I am not unlike a detective who works his way backward using the evidence. Granted, I attempt to reconstruct events that may have happened thousands of years ago, and instead of examining hours-old corpses and ancient fossils, I look to the living for evidence of genetic events that have come to characterize the development and behavior of certain species.

When I mention genes, images of scientists in lab coats, microscopes, slides, and tiny glass vials may come to mind. But there are other ways of gaining insights into existence. Provided one has a healthy sense of curiosity and a basic understanding of natural selection, one can make significant inroads in the search for self and other knowledge. Natural selection refers to the long-term process whereby survival-enhancing genes are selected over those genes that do not impart fitness benefits to their bearer. We might think of ourselves and other creatures as survival machines that have taken thousands, if not millions, of years to refine. When we look at our own bodies and behavior and at the other organisms with which we share our lives, we are seeing the culmination of this process. Suddenly, the questions "How old are you?" and "Where are you from?" aren't so easy to answer.

Keeping in mind the complex co-evolution of genes (the biological units of transmission) and memes (the cultural units of transmission), we can also look to our cultures if we wish to gain insights into human nature. Human cultures have some important differences, but generally they are uniform due to their biological underpinnings. For example, human and other primate societies generally have strict taboos against incest, but it is also true that reproducing with closely related DNA is detrimental to genetic integrity and, ultimately, to genetic survival. In fact, primate studies have shown that siblings who are raised together for the first thirty months of life develop a sexual aversion to one another, suggesting what sociobiologists describe as an epigenetic rule or, in this case, a rule of inhibition.

It is this complementary relationship between genes and memes (the nature v. nurture dichotomy) that makes it possible to explore both the biological and the environmental implications of our stories, especially as they are expressed through and reflected by the arts and other human activities. In the chapter "Arachnophilia," I spend considerable time attending to the lives of several species of spider with which I share my yard, but I also lament how the stories told by our culture have really done a disservice to these remarkable creatures. Clearly, it's always wise (and natural) to be mindful of potentially dangerous creatures (I can think of only a handful of spiders in North America that pose a significant threat to humans), but there is a crucial distinction between being mindful of and unnecessarily hostile toward other creatures which are, after all, simply trying to live their lives.

Over the course of three months, I studied a black widow that had made a home on the east side of my house. I learned a lot from her, but one of the more poignant realizations I had was how fundamentally similar all animals are in their day-to-day lives. Most of us never get beyond our differences of appearance, but if we did, we would see that our basic needs for food, safety, and shelter are essentially the same. How different species go about fulfilling their basic needs for resources and shelter is amazing. Habitat theory, which I explore in "A Field Guide to Habitat Theory," argues that animals select habitats based on how well they fulfill their biological needs for prospects and refuge.

For example, we have a mature orange tree in our backyard, and for the last week or so a pair of mockingbirds has been busy constructing a nest in the tree's inner reaches. In the three years we have lived here, I have never seen any birds nest in the tree. Our two cats patrol the yard, so I've always assumed that the birds opted not to nest close to predators. But here these birds were, building away. I wondered what had changed. The answer was suggested to me by another bird, the curve-billed thrasher. Each afternoon, a pair of thrashers would spend an hour or so rummaging through the leaf litter and rocks beneath the orange tree. I quickly realized what they were looking for: we have two box turtles and the female had laid her eggs beneath the tree. I don't know if the thrashers were actually digging down to the eggs themselves, but I do know that they and other birds enjoy the mucus plumes that signal the location of the eggs.

Many species of bird have their chicks in the spring, a time when protein is in high demand. Animals must balance their needs for resources and safety, so I'm guessing that, for the mockingbirds, the benefit of the resource outweighed the apparent cost or risk of nesting so close to predators. I imagine that living above the turtle protein would be a little like living above a butcher shop. All similes aside, animals—birds, humans, and even black widows—are generally attracted to habitats that put them in proximity to resources while minimizing risks to themselves and to their offspring.

One of the reasons habitat theory is so interesting and useful from a human perspective is that it informs our ideas about beauty and aesthetics. In addition to speculating about how the theory might explain my own habitat selection and, as it turns out, creation (I've spent hours beautifying our yard), I also cite recent research, including exhaustive studies of landscape paintings, which suggests a biological—as opposed to a strictly cultural—basis for our aesthetics. While environments can evoke complex and sometimes conflicting psychological responses, the studies and my own research show we generally find beautiful those environments which signal prospects and refuge, whereas we usually have negative responses to environments that do not. This helps to explain why the landscape painter Thomas Kinkade is a multi-millionaire, as well as why people who are unfamiliar with deserts often have difficulty thinking of them as beautiful.

In the final part of the book, "Implications of an Eco-childhood," I reflect on my experience growing up in the deep woods of northern Maine and, later, in the high deserts of Utah, where I spent hours engaged in various outdoor activities, including hide-and-seek, fort building, stick fighting, and hunting. I don't think anyone would argue with the notion that domestic kittens are predisposed to stalk one another because it helps them develop the skills necessary to procure their own food later in life. To the extent a kitten's interest in stalking helps to ensure she will reach reproductive age, the behavior might be described as adaptive. I began to wonder if my own and, by extension, other children's play interests are also adaptive. If so, this might give us a richer understanding of play's importance and complexity, as well of the need to preserve and create sites for that play.

Playing tag and rock fighting might both be described as adaptive behaviors, but obviously we encourage our children to play tag, whereas we forbid them to rock fight. That said, we risk missing the point altogether when we label these behaviors (or kids) "good" or "bad." Our ability to effectively address these tendencies among children and ourselves is contingent on our understanding of those tendencies. That's another thing I love about Darwinian evolution: used honestly, it serves everyone or no one; includes everything or nothing. It embraces our finest moments and our worst; it is awareness in the deepest and truest sense. The question is do we have the courage to seek out this knowledge and the wisdom to live by it once it's found?

ARACHNOPHILIA

September has come to the Sonoran Desert. Now that the mornings have begun to cool, I like to wake an hour or so before dawn, walk through the house, and open all the windows. The cats follow me out of the bedroom and then race ahead as we near the patio door. They are eager to go outside to investigate their territory and to lie belly down in the cool grass. Every morning our routine is the same: I perform my rituals in just this way, in just this order, and rarely deviate. Because of my early rising, I have time to wander in my yard to see what new things have happened in the night. One of the creatures I've been keeping tabs on is a large, lovely black widow spider (*Latrodectus hesperus*) on the east side of my house.

Around this time last year I counted four males and eight females, their webs on all sides of the house but especially on the west side. This morning I can find only one. She, too, has her rituals, albeit they appear less numerous than mine. Each evening, just as the yellow-blue sky of dusk turns to the gray-purple sky of dark, she appears in her web, where she settles in for a roughly twelve-hour wait for food. Now it's 5:30 a.m. and she's still out, inconspicuous to all except those who seek her or, in the case of the many crickets and cockroaches that scuttle along the cement slab below her web, those who inadvertently find her.

Since she is most active at night, I often check on her three or four times before turning in for the evening. I find her hanging upside-down in the same spot each time, poised to engage with remarkable speed any insect that happens to stumble into her web. I note the deep red hourglass on her abdomen in full display and it occurs to me that I have never seen a black widow in any other position—that is, unless she is repairing her web (I have not yet had the pleasure of observing the notorious mating ritual that sometimes ends with the cannibalism of the male by his mate).

That she would adopt and maintain her resting position while lying in wait for prey makes sense to me for at least two reasons. Compared to the webs of orb weavers, which are constructed with a geometrical precision that inspires awe and amazement, at first glance the webs of black widows appear chaotic, a tangled mesh of intersecting strands haphazardly thrown together. On closer inspection, however, I see that a black widow's web has a certain structural logic, supported by the spider's preference for crickets and her predatory position within the web.

Comparatively speaking, black widow webs are easy to find. In the two years I've lived here, I've seen roughly fifteen webs around my house, every one of them in similar locations. Based on my observations, widows prefer to be close to the ground and close to their prey. What I find especially interesting, however, is that all the webs were constructed adjacent to concrete and were heavily concentrated over open ground. The operative word here is "concrete," for, unlike the 3/8-inch gravel spread between the house and walkways and driveway, concrete provides a stable surface for a web. If the web is going to be effective, it's got to withstand the spider's movement and the force of struggling prey. I can view the web builder easily, especially if the web is not too cluttered with bougainvillea leaves and other wind-blown refuse.

We can tell precious little about how our human neighbors live or what goes on inside their dwellings: here in the West, at least, we spend much of our lives out of view. It is just the opposite with black widows. Although the cover of dark provides concealment, how these spiders live their lives is readily discovered with the aid of a flashlight. When I go out into the yard at night, flashlight in hand, and see the spider poised in her web, I can appreciate the logic of the web as it relates to her needs. The size of the black widow's web doesn't help me understand the web's connection to her prey preference, nor why she would be thus positioned in anticipation of wandering crickets. But the strength of the web is another matter.

Figure 1: Black widow

A black widow's web will accommodate insects of various sizes. Just last night, I observed a tiny ant of an unknown species caught in the widow's web. A brief struggle ensued, during which the ant must have nipped the spider because suddenly the predator became agitated and retreated into the ruellia bush. But the toughness of the web's silk far exceeds what is needed for ants. I imagine that the web helps the widow help herself to fairly large, potentially harmful creatures, such as crickets and cockroaches. As my examination with a handheld magnifying glass revealed, both species are equipped with mouthparts that could seriously damage the black widow. Perhaps for this reason the spider possesses an extremely powerful poison: she may only have one chance to bite and neutralize her prey, so it better count. But she seems to have other protective strategies as well.

I suppose that is the whole point: her anatomy is in part the result of her prey's anatomy, especially its armaments. For one thing, her legs are long and slender relative to the heft of her robust abdomen. What her legs lack in bulk, however, they offset with length, an appropriate adjustment—her appendages must not only support her weight, but also enable her to move nimbly within her web. That her legs are geared toward facilitating movement within the web (which she rarely leaves) is also suggested by her comparatively awkward movement along the ground, where she appears to bounce slightly as though on springs. Equally important, however, is how the widow's legs assist her in subduing prey. On the two occasions I have seen her capture prey without any help from me, she began by wrapping the victim in the web before administering her first bite. When the prey was a cockroach, she wrapped the head first, incapacitating the cutting mouthparts. When the prey was a cricket, however, she first immobilized the insect's rear legs by wrapping them in silk.

A cricket's rear legs are very strong; they would have to be in order to propel the leaping insect up to twenty times its body length, but they are also heavily spurred and might enable the cricket to free itself or fend off its attacker. Either outcome would be undesirable, perhaps even fatal, to the spider. In order to deal with these potential problems, the spider has developed long slender legs that enable her to pay out and apply silk while keeping her body at a safe distance from her victim. Of particular importance here are the stiletto-like outer segments of her legs, which appear to be less than half the diameter of the upper leg, and would seem to offer less of a target for struggling prey.

I can also see how the black widow's anatomy in general and thorax-abdomen relationship in particular might have been influenced by predation from birds and other animals that attack from above. Otherwise, the widow wouldn't have adopted her upside-down predatory pose, which allows her to display her telltale marking to potential predators, at least those that attack from above, while maintaining a functional relationship to her web.

One night, I was out on the back patio and noticed that a widow had taken up residence under our Ping-Pong table. Her web was the usual array of mesh and long strands, but unlike the spider on the east side of the house, whose web was made in a low-lying ruellia bush, this female was suspended roughly two feet above the ground and, therefore, her intended prey. Perhaps she compensated for this apparent handicap by making a larger web, one with a volume four times that of the ruellia bush spider. Still, the ruellia bush web was much closer to ground, and so would seem to offer more ready access to prey despite its smaller size. In any event, after a few minutes, the Ping-Pong table widow began lowering herself to the ground, where she tapped her spinneret at the tip of her abdomen, and then hoisted herself high into the web. She repeated this process for about five minutes.

The patio light prevented me from getting a good look at her handiwork, so I squatted down: now I could see what she had done. Descending vertically from the mesh were long strands of silk or "capture fibers" that stood alone or were crossed by other strands, and these appeared to be the primary mechanism for capturing prey. With the help of Ken and Rod Preston-Mafham's book *The Natural History of Spiders*, I later learned that web-building spiders use different glands to manufacture several types of silk, each designed for a specific purpose. The spider had mesh around her lair, into which she would withdraw during the day or when threatened, but this part of the web, also known as "scaffolding," is not responsible for the capture of prey.

Later that night, I returned just in time to see the Ping-Pong black widow capture a cricket. The cricket was traversing the patio when, I assume, it made contact with one of the capture fibers. I couldn't tell what part of the cricket first made contact with the web, but judging by the length of the cricket's antennae, it surely tapped the web first. At that moment, the widow dropped like a stone from the mesh and began ensnaring her victim. What struck me most was how fast she descended—in the time it takes to blink, she had traveled the two feet and snagged her prey. Given the precision with which she located the cricket, I suspect she actually raced down the web, keeping contact with at least one silk strand as she descended.

After consulting my *Audubon Society Field Guide to North American Insects and Spiders,* I learned that the black widow has three claws on the end of each tarsus (the Kung Fu gripping device of the spider kingdom). She uses these claws to hold on to her web and, as I saw last night, to carry prey into her lair, which she accomplishes by grasping the victim with one of her back legs and trailing it behind her. She may carry other items as well, such as leaf litter and insect offal. Like other web builders, the widow maintains a close relationship with her web, but some spiders dive out of their webs when they feel threatened. For example, once there was a species of orb weaver living in the hibiscus plant on the east side of my house. The spider had built its web toward the center of the plant. When disturbed, the spider free-fell (similar to bungee-jumping, but without the recreational component) about ten inches into the center of the hibiscus plant, and was removed from view.

As these insights into the lives of the widow and the orb weaver suggest, spiders of all kinds have developed sophisticated survival strategies—but instead of being portrayed as complex animals that act sensibly according to the day-to-day challenges they face, spiders are often depicted as deeply malicious creatures, hell-bent on finding, pouncing on, and biting unsuspecting and innocent humans. It is a wonder to me that I, too, did not succumb to an unchecked fear of these interesting and beautiful creatures, for I recall seeing several spider horror movies as a child.

Actually, giant spiders from outer space weren't the only threat to the townspeople of Caribou, Maine, my family, and me. While I lived in northern Maine from 1968 to 1979, roughly the first eleven years of my life, several "natural horror" films came to the single-screen movie house down the street. I remember four of them clearly: *Grizzly* (1976), *Day of the Animals* (1977), *Orca* (1977), and, most terrifying of them all, *Prophecy* (1979), which had the added impact of being filmed in northern Maine. Each of these movies attempted to address the problem of what happens when humans encroach on, fail to respect, or in some way "wrong" other animals or the environment.

The animals in question were not depicted as the mammalian equivalents of the malevolent great white shark in *Jaws*, which came out in 1974. They did not kill for the sake of killing: they killed because of human misconduct, for a reason other than blind hatred. This made them more terrifying to me, not less. It is hard to say what effect these movies had on the general public, but for a child living in the deep woods of northern Maine who was a three hour drive from the shark-filled Atlantic Ocean, they offered compelling reasons to revere and respect other animals.

The fear I felt as a result of these films did not make me hate other creatures. Rather, the films alerted me to the consequences of failing to respect other animals and their homes. This may not have been the case for other children (and adults) who saw (and continue to see) movies such as these. I viewed these "natural horrors" in the context of my generally positive experiences with real animals. Contrary to what the directors may have intended, I saw that the true "natural horror" was actually the juggernaut of human processes that destroyed the ability of other animals to live their lives free from constant interference.

Perhaps the prime natural horror film was *Twister*, which appeared in the late nineties. In this movie there are several chase scenes; sometimes the tornado hunters do the chasing, but they also spend a good deal of the time being chased by the tornados. The idea that a tornado is actually doing the chasing reveals the human tendency to assign ill-will to natural phenomena we don't understand. The real kicker, however, came during the final chase scene; the grand-daddy of all tornados was bearing down on the "scientists," and as it drew within striking distance, above the cacophony of wind blowing, corn lashing, thunder pounding, and people screaming, the tornado let out a deep and unmistakable growl, as though it were some great beast come to quash the innocent, if hubris-afflicted, humans.

I don't mean to belabor the point, but I think movies like these often perpetuate the paranoid notion that nature has consciously destructive designs on us. Moreover, they suggest that nonhuman nature is impervious to human attempts to understand it, and that if we do attempt (for example) to penetrate the tornado's innermost reaches and learn its secrets, it will retaliate by destroying us. Obviously, tornados don't have any desire to harm us, just as the overwhelming majority of large carnivores and poisonous creatures do their best to avoid us, rather than attack. Knowing this, I take great care not to flash my light directly on the black widow as she attends to a cricket she has captured, nor do I so much as breathe on her or her glistening web.

I've always been curious about other creatures, but the intensity of that curiosity, and the lengths to which I will go to satisfy it, depends on the species. Even now I am, like most people, more interested in some animals than in others, but I am coming to realize that each and every creature might—if given adequate time and study—reveal a complexity deserving of my attention. The real question, then, is not whether I value one creature more than another, but rather which creatures will I notice and why. Perhaps every creature warrants our attention and understanding (although I confess I attend to mosquitoes by killing them). Considering there are literally millions of organisms, many of which have been studied very little if at all, the task ahead is daunting if we are to achieve even a rudimentary understanding of all creatures great and small.

Insects, spiders, and the small things of the world often escape our notice because they are trying to hide from their predators, which they accomplish by blending in with their environment. The black widow's nocturnal activity is almost certainly a strategy for minimizing the threat of predation by birds. But predator and prey co-evolve, so although hunting at night may lessen the threat, it does not dispense with it. This may help to explain the widow's pitch-black coloration: against the backdrop of night, she is imperceptible. There is, however, one problem with this line of speculation: it does not explain why the widow often emerges from her lair at dusk, a time when she is still highly visible.

I have noticed that, generally, my favorite black widow stays within an inch or so of her lair from the time she emerges at dusk until it is fully dark. As night deepens, she moves farther out into the web and away from her retreat. If this behavior is adaptive, then the benefits of emerging before nightfall should outweigh the risks. Many of the widow's predators—including other spiders—are nocturnal. The other major predators of the spider, birds, roost in the evening and have all but disappeared by dusk. If anything, then, pressure from nocturnal predators would seem to be less during dusk than dark, if only because the hunt would not yet be in full swing. Perhaps dusk presents a relatively safe time for the widow, a period when she can hunt with reduced risk. If I am right, I would expect to see crickets emerge around the same time. Last night I didn't see any crickets until full dark, right around 7:00 p.m. But I did hear them as early as 5:30 p.m. Perhaps the widow emerges soon after the time male crickets start singing because that is when the females start wandering in search of them. Two weeks of data would appear to support this hypothesis—the black widow usually appears during the hour or so between the time male crickets start singing and female crickets begin wandering.

Dusk or dawn, night or day, the most obvious way most

insects and spiders avoid human detection is to be, by virtue of their small size, anything but obvious. I have 20/20 vision, but as I stand above a litany of ants making its way across the patio, I've got to concentrate in order to register what I'm seeing. Being bipedal and of average height, the visual command center in my head is physically and spatially removed from the world of small things. Therefore, unless an insect scuttles across my field of view, I won't notice it.

When I look back over the last two years, I see my interests have alternated between the macro features of my yard—such as trees and bushes and their arrangement—to the micro features, including insects and spiders. I am reminded of the all-encompassing excitement I felt when Kim and I moved into our new home on Halloween. We didn't get much done that first night: every five minutes or so, the doorbell would ring and there would be children in various costumes holding out their pillowcases and paper bags, asking for candy. In the days that followed, I often neglected the post-moving minutiae, such as unpacking and shelving my books. Instead, I favored wandering out into the backyard to plan for its renovation.

Much to our delight, the landscaping had all been done six years before, so the trees and bushes were already well established. The east side of the house—a beautiful area with robust bougainvillea growing along the north wall, a grapefruit tree to the east, and a Chinese elm tree in the center—is my favorite side, but at that time it lacked ground cover, and so promised to become a dustbowl or a muddy quagmire, depending on the weather. As I stood out there one November morning, trying to calculate how much sandstone I would need to cover the area, I noticed a spider web between a stack of patio chairs and the air conditioner. I bent down to take a closer look: the web was a perfect orb, undulating like a medusa jellyfish in a sea of air, quite unlike the seemingly chaotic web of the black widow.

The resident spider was nowhere to be seen. What I could

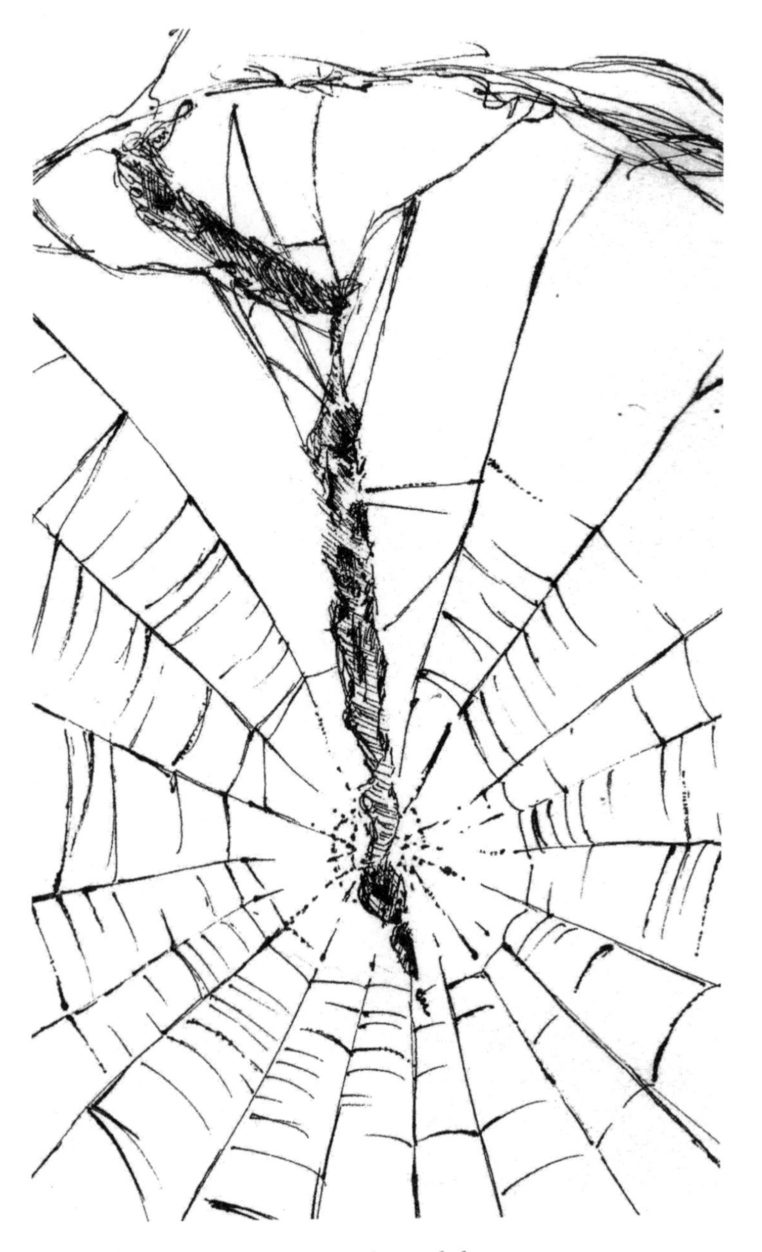

Figure 2: Araneidae stabilimentum

see was a litter of bug husks and other organic matter lashed together with a light mesh of silk. This material formed a vertical line that extended two inches in either direction from the center of the web. I would later learn that this structure is called a *stabilimentum*, and that it has several possible functions, including providing shelter for the spider. Only after examining each segment of this curious structure did I finally detect the spider in her lair. She had made a tiny, silken tent in the very center of the web, into which she fit like a key. A little larger than a peppercorn, her abdomen fit snugly into the lair, where it was hidden from the sun and the searching eyes of predators. She was not the most visually striking spider; her pale brown abdomen was lightly marked with what resembled a stretched triangle. In fact, were it not for the stabilimentum, I may not have noticed her at all. I do not begrudge her that. After all, with so little between her and an array of predators, it's best to keep a low profile.

The success of an orb weaver's web depends on its stickiness and flexibility, but its silken lines must not be so dense that they alert prey to danger or attract predators looking for a snack. Then why make a stabilimentum that draws attention to an otherwise nearly invisible web? Returning to *The Natural History of Spiders*, I read that some have speculated that the stabilimentum alerts birds so they do not fly into and destroy the web. On some level, this would seem to defeat the purpose—an alerted bird is a deadly bird. Assuming this explanation is plausible, it probably applies only to spiders that construct webs in areas where birds fly, such as between trees. But the side of my house, next to an air conditioner and a stack of chairs, would not qualify as a major flight corridor. If anything, given the web's location, the spider should do whatever she can to make the web *less* apparent, not more. So what's going here? Given its prominent position within the web, the stabilimentum would seem to have some structural, perhaps unifying, function in addition to concealing and sheltering the spider. But there is also at least one other explanation, which I learned when reading about a group of spiders in the genus *Argiope*.

Many spiders of this genus construct stabilimenta that are so conspicuous that they make my spider's stabilimentum seem positively ho-hum. There are two species of *Argiope* found in North America, a black-and-yellow species and a silver one. Although their stabilimenta aren't as eye-catching as those of some tropical species, they are still far from discrete. While some drab orb weavers use insect remains to construct their stabilimenta, species of *Argiope* use a special silk to create dense zigzags or discs of shiny web. Because many of these species build their webs in forests which have low light levels and are in close proximity to avian flight corridors, perhaps the stabilimentum alerts birds and prevents them from colliding with the web and the spider sitting in the center of it. But this conclusion still does not account for the startlingly conspicuous nature of the web, which would seem to attract predators in search of a chunky spider to eat.

The most compelling explanation, though not without its problems, is that stabilimenta are used to attract insect prey to the waiting *Argiope*. Some spiders accomplish this end by using silk that reflects ultraviolet light, which, from an insect's point of view, mimics flowers' ultraviolet reflections. It's unclear, however, as to whether the stabilimenta of all spiders reflect ultraviolet light, or if they employ some other means, such as a chemical attractant, to lure prey to the web. It's also possible they use a combination of strategies.

Perhaps the shapes of the different stabilimenta offer insight into how they appeal to prey. The stabilimenta of juvenile and adult *Argiope* are quite different; youngsters construct a disc while adults construct a cross of zigzagging ribbons. But both are significantly more ornate than the stabilimenta of the *Araneidae* orb weavers I've seen, including the one that made its home on the east side of my house. To me, the stabilimenta

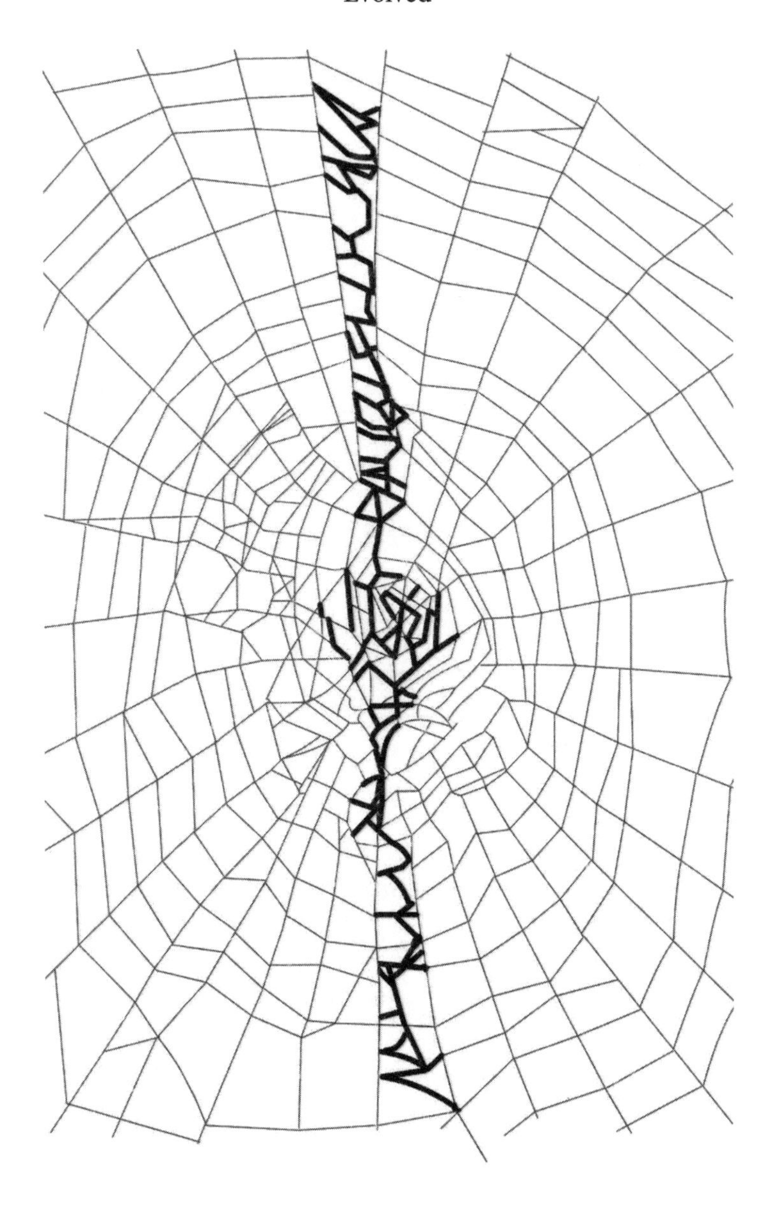

Figure 3: Argiope stabilimentum

of *Argiope* actually look like flowers, whereas the *Araneidae* stabilimenta do not—they look like body parts dangling from a trotline.

In either case, the plausibility of this "fatal attraction" theory hinges on the notion that by increasing the capture rate of prey, the spider increases her reproductive success. This makes me wonder how the spider ensures that she herself will not be preyed upon before she has a chance to reproduce, as well as how the stabilimentum compensates for the potential hazards of being so conspicuous. As with other adaptations, purpose depends on the audience, in which case an adaptation may simultaneously function on several levels. For example, the bright coloration of some insects serves as both ornament (by suggesting health) and armament (by suggesting toxicity). Considering the all-important relationship a spider has with its web, which acts as an extension of the spider, the stabilimentum may also function as an offensive mechanism or armament.

Keeping audience in mind, perhaps the stabilimenta of some spiders attracts or repels depending on who—prey or predator—is expressing interest. Maybe stabilimenta supply odors that attract prey. The stabilimentum might also dissuade predators from investigating by mimicking bird droppings or some other form of inedible detritus. Some spiders have evolved resting positions that mimic bird droppings, so this hypothesis may be plausible. In the end, considering the orb weaver seems at even greater risk without the stabilimentum, I wonder if the stabilimentum is really all that puzzling. After all, it provides shelter and protection, as well as a means of attracting prey. Many spiders forego the "bells and whistles" of adornment and use only their webs to meet the goal of survival.

Clearly, this little orb weaver on the side of my house had a lot going for it, including protection (and the occasional feeding) from me. I remember one evening when Kim's parents came over for dinner. It was breezy with high clouds and the promise of a spectacular Arizona sunset, so after we had eaten, the four of us decided to have our coffee outside. I didn't notice that my father-in-law had walked around to the east side of the house to get a couple of patio chairs. He hadn't been gone for more than a few seconds when I realized what he was doing and ran to the east side of the yard. Just as I rounded the corner, he was placing his hands on the armrest of the patio chair, preparing to lift it from the stack. "Stop!" I said, not quite yelling. Don looked a little startled, that childlike *What did I do wrong?* expression on his face. I quickly explained that a spider had made her web between the chairs and air conditioner. "I've been kind of looking out for her," I added, just in case he was feeling slighted. "Oh," he said, as if caring for the spider made all the sense in the world.

Given the underlying fear many humans have of spiders, my concern for the orb weaver's well-being puts me in a distant minority. Even though, over the course of our evolution, we developed a fear of spiders and other potentially dangerous creatures, we also developed complementary abilities that would, under the right circumstances, allow us to check our fear in the interest of satisfying our curiosity. This may help to explain the simultaneous aversion and fascination primates feel when encountering snakes and other poisonous creatures.

The extent of human curiosity and therefore the degree to which we can learn about other creatures obviously varies from individual to individual. Despite, and at times because of, the culture in which they occurred, my own life experiences have moderated my biological predispositions, including a fear of spiders. But even if I were a confirmed spider-hater, it's hard for me to believe I wouldn't have been awestruck by the intricacy and beauty of the orb weaver's web. To be awed requires a close look, and I cannot remember having really looked at one until that day I first glimpsed the silken disc on the east side of my house. Over the following weeks, I spent many hours observing the spider, and with each new bit of information about her world I became increasingly attached to her.

Early in my relationship with the spider, I would sometimes feed her a lacewing that I had captured by rousting them from the branches of our orange tree. Invariably, several of the delicate insects would rise out of the leaves, fly across the yard, and alight on the side of the house. I would gently pinch their wings between my fingers and deliver them to the spider, which I did despite a small dollop of guilt that I felt in my role as executioner. I eventually stopped playing "meals-on-wheels" with my spider, but I wondered why it took me so long to decide that feeding lacewings to the spider was morally suspect. On the one hand, I felt the need to respect the lacewing as another life form, but then there was the competing and apparently stronger need to understand the orb weaver, which I did by feeding her. Fundamentally, there is no way to justify my treatment of the lacewings, but part of me wants to excuse my behavior on the grounds that I did it to educate myself about spiders.

Beyond a certain point, the extent to which we know other creatures determines whether or not we will include them within our moral framework. We often have difficulty appreciating other creatures because of our perceived lack of access. The notion that nature is "out there" is true to the extent that certain environments, such as national parks and wilderness areas, have been set aside to be kept intact. This is important for the preservation of these places, but it also tends to obscure the awareness that nature is everywhere. Oddly, even though our immediate awareness of our natural surroundings is often overshadowed by our dreams, imaginings, and trips to far off places, our survival depended (and still depends) on how well we came to know our immediate environment, including other animals.

Once I learned that lacewings prey on destructive insects,

I no longer fed them to the orb weaver. I also began to appreciate their beauty, especially when in flight. This wasn't an instant awareness, however. It was gradual and cumulative. Long before these insights finally prompted an end to my behavior, I struggled with the moral implications of interfering with the natural scheme of things. I realize that my language here implies that I am somehow outside nature and that nature is separate from me. But my language is misleading. Regardless of how unsavory, my willingness to interfere with the lives of the lacewing and the orb weaver is no less a part of my nature than is my impulse to protect the orb weaver from my father-in-law. Still, however natural it is for me to feed the orb weaver lacewings in order to sate my curiosity, it is not desirable from the standpoint of respect and noninterference.

Toward the end of my service to the spider, I would often play a game with the lacewings that illustrates this conflict between my desire to watch the spider kill and my awareness that lacewings deserve to live too. One of the problems I encountered when trying to feed the spider was that the lacewing sometimes wouldn't stick to the web. The lacewing would fly out of reach, or would alight within my reach, apparently unaware of the danger I still posed. I actually preferred it when they flew up and out of reach because it meant I wouldn't be faced with recapturing them and trying to reinsert them into the web. As time went on, and I learned more and more about the lacewing, I abandoned this indiscriminate approach in favor of what I considered an arguably more honorable approach. Assuming the lacewing didn't get caught in the web the first time around, but had alighted nearby, in the spirit of fair play (I hear the lacewings say, "Don't do us any favors, jerk."), I agreed not to recapture the lacewing and make another attempt to feed it the spider. If they escaped, they were free.

The new rule of my game offered a false compromise because it merely enabled me to resolve conflicts within myself while still subjecting the lacewing to a fifty/fifty chance of death. The fact is, had I not disturbed it and captured it in the first place, there would be no need for a compromise. In any case, my actions may be perceived as more or less problematic depending on whether and how I think about them. What would happen, for instance, if instead of working through my feelings of guilt, I had secreted them away or ignored them? Human emotions and our encounters with the nonhuman world have this in common: if approached openly and honestly, they reveal the depth and richness of life and experience, however dark and conflicted they might be.

One of the most important lessons I learned during my study of the orb weaver is that I cannot take for granted the accuracy of my thoughts, nor any conclusions I may draw from them. This became especially clear to me one morning as I attempted to feed the orb weaver a lacewing. The spider was nestled in her silken tent. Although her abdomen was obscured from view, I could see her front legs touching the thick radial lines on either side of her tent. The radial lines connect the spider to even the subtlest event in her web—they are like a motion-activated alarm system. The spider had been busy in the night repairing her web. Her disc was whole all the way around, free from tears and breaches that the lacewing might exploit to free itself. Consequently, the lacewing quickly became entangled once I threw it into the web. It tried to free itself, its jaws cutting wildly at the silk, but to no effect. Then something extraordinary happened.

The entire web shuddered as though it had been shocked. Initially I thought it was a fluke, or the wind, or a trick of the eyes, but then it happened again, and again. I had seen this while watching a species of marbled spider, dozens of which have made their homes in the rosemary bushes on either side of the orange tree. But the shuddering of the orb weaver's web was different. For one thing, whenever I had observed the marbled spider, the web didn't shudder so much as the spider itself did. In fact, the spider did more than shudder: it bounced so rapidly that I had difficulty focusing on it. It's possible that the marbled spider shakes the web in order to further entangle prey, but considering I'm not prey and that no prey was present when I observed the shaking, I think the bouncing was a defensive strategy: if I can't focus on the spider, there's a good chance I won't be able to pluck it from the web.

In the case of the orb weaver, however, prey was present. Did this mean that the spider shook the web to further entangle the struggling lacewing? Perhaps. But I think there was something else going on, too. I wondered if maybe the orb weaver was, like the marbled spider, shaking her web as a defense against me. I dismissed that interpretation, however, because the orb weaver's web shook while she herself remained relatively still. Also, the orb weaver has her tent to conceal her from view, whereas the marbled spider has no such shelter and instead retreats into the rosemary or employs the bouncing technique described earlier. For a moment I also considered the possibility that maybe the lacewing was too big and the orb weaver was shaking the web in an effort to dislodge the lacewing before it could cause further damage. I junked this hypothesis because it just wasn't logical. For if the lacewing were too big, it would have freed itself—that is, assuming it got snagged in the first place.

Although logically implausible, this last hypothesis differs from the others in at least one important respect: it attempts to understand the event from the perspective of the participants, rather than from my perspective as an observer. The moment I began to ask questions from the perspective of the spider *and* the lacewing, rather than from my own point of view, it occurred to me that the shuddering may have been a defense mechanism against the lacewing, and a violent one at that. All things being equal, lacewings can take care of themselves. They are highly effective predators, and their powerful jaws might harm the orb weaver. By shaking the web, the orb weaver may further entangle the lacewing, but perhaps the more important result is that she disorients it, which makes it safer for her to approach.

As a rule, we are biologically conservative, and though we may have moments of insight into the bigger picture, they are the exceptions. These insights usually only come with great effort. It appears the human mind, which evolved to keep us alive, must be trained to fathom the world from an ecocentric point of view. The essential difference between the ecocentric and the anthropocentric view of other animals is that the former values other creatures apart from their roles as prey or predator. Certainly I am capable of appreciating the complexity of animal behavior, but I am not prone to it. Obtaining food and avoiding predators have been major preoccupations for roughly 99 percent of human history, so it seems our thought patterns and learning biases would reflect this specialized relationship.

Therefore, despite my wishes to the contrary, I have to constantly remind myself to think about how I am thinking and to ask whether or not I understand what I am seeing. Obviously, I could ask this question and still be at a loss, especially if I didn't have recourse to an alternative thinking technique that would enable me to recognize and surpass my intellectual shortcomings. This is eventually what happened after I had had an encounter with a jumping spider in my front yard. We have a planter on the west side of the yard that attracts insects and spiders, as well as serves as a nursery for baby lizards. The box is filled with various flowers, but at one time it doubled as a litter box for the many cats—feral and domestic—that roam our neighborhood. Because of the flowers and the feces, the planter is a bug haven. And where there are bugs, there are the animals that prey on them.

When I am still, I sometimes see jumping spiders in the grass, jumping from blade to blade in search of prey and mates. Against that background, they are imperceptible unless they move. The planter, however, is white and contrasts sharply with anything that is not. One afternoon when I was using an electric weed-eater to trim the grass around the planter, I saw a dark-bodied jumping spider traversing the white planter. Unlike orb weavers and other web-spinning spiders, jumping spiders rely on their vision to detect prey. They have four sets of eyes, each of which has a specific function. The spider's anterior median eyes are located front and center of the spider's head, and they are the largest and only closely set pair. Because of their size and position, coupled with the way the spider tilts up its carapace and thorax once it has detected me, these eyes have on many occasions given me the unmistakable feeling of being looked at by these tiny, inquisitive spiders.

As I watched the spider, I noticed he would hop and begin waving what I later learned were his palps, which are two leg-like appendages located on either side of the spider's mouthparts or chelicerae. I was completely baffled and entranced by what I was seeing. Watching the spider's palps quiver and undulate reminded me of watching someone dance on a disco dance floor under the halting and fracturing effect of a strobe light.

As a burgeoning naturalist, I began to ask questions about the jumping spider's palpitating palps. Strobe lights have several effects, and distortion is certainly one of them. I felt confused by the spider's movements. I wondered if the movements were actually a sophisticated defense mechanism used to confuse and discourage me, a potential predator, from attacking. This hypothesis seemed more plausible until, a couple of minutes later, another, much larger jumping spider suddenly appeared on the planter seemingly out of nowhere. I assumed the larger spider was an intruding male, and that the smaller spider was moving his palps as a threat display. Therefore, as far as I was concerned, my initial hypothesis remained intact except for one minor adjustment: the palps were used as a defense not against me, but against other spiders.

A few months later, I attended a lecture by a professor who, although her talk happened to be about Darwin's discussion of seed dispersal, I knew was a spider expert. After she had finished her talk, she lingered to take questions, which was great because I had some questions in mind. Unfortunately, an elderly man with a guitar over his shoulder had gotten to her first. I waited patiently as the gentleman asked her questions about God. I could tell the conversation was taking its toll on the professor, so I took a step closer, at which point she turned to me and, pleadingly, invited me to join the discussion. I apologized to them both and explained my question was not about God but something a little closer to home, namely, jumping spiders. Not surprisingly, the professor beamed. (Exit her evangelical questioner, stage right.)

I told her about my encounter with the jumping spider, and then, feeling rather confident in my powers of observation, I shared my defense hypothesis. As I explained my theory, her eyes widened and she smiled—I could tell she was excited by my findings. Turns out she was excited, not because I was right, but because she knew what I had really seen, which was most likely a courtship between a male and a female. She pointed out that, while their role varies from species to species, palps are indeed used by the jumping spider as display devices for warning other males, but given my observations, the palps were being used as ornaments for attracting and later inseminating the female. The complexity of the palps depends on the spider, but in general the palps are penile in function. Different spiders acquire and inject their sperm in various ways. Forgive the comparison, but the palps of the jumping spider are similar to turkey basters in the sense that they can suck up, carry, and squirt out sperm: talk about multi-tasking!

In retrospect, there are a few things about my encounter with the jumping spiders that reveal my limitations as an observer. For instance, I initially didn't notice the larger female spider. If what the professor said is true, then the female was within view of the courting male spider, which is why he started his courtship routine. Otherwise, the male would not have gone through all the effort. If the female were within the male's field of view, and I could see the male, why didn't I notice the female when she was a mere ten inches away? Because I was focused on the male, it was easy to overlook other objects in the vicinity, especially if they were motionless. In addition, I automatically assumed that the spider's movements were directed toward me, in which case there was no motivation for me to consider alternative explanations. A compounding factor was my assumption that the male spider's movements were aggressive, which again would have kept me from looking for a sexual partner rather than an opponent or enemy.

My assumption that the male spider was acting aggressively may signal a learning bias. This tendency would be less acute during encounters with low-risk animals and more acute in interactions involving poisonous or otherwise dangerous creatures. In the ancestral environment, traits may have been selected that heightened our awareness of and caution toward certain creatures, while other traits with the opposite effect may have been selected when the creature was innocuous or had low-threat relevance. When I consider my own heightened state of caution-inspired awareness (in the company of the black widow, for instance), and then contrast it with my relaxed state (in the company of a lacewing), it seems clear that the former would be much more exacting in terms of energy expenditure.

I wonder what would happen, then, if we were unable to differentiate our encounters with other creatures and were thereby unable to respond appropriately. Living in a state of hyper-awareness would surely have been unsupportable, but living in a state of hypo-awareness would have been equally costly. Our responses prompt us to conserve or expend energy depending on what the situation demands. This helps explain why I was quick to conclude that the male jumping spider was acting aggressively, a perception that triggered a cautionary reaction from me. There may be other factors as well, such as my assumptions about the spiders and their size difference.

In general, human males and the males of other primate species tend to be larger and more violent than their female counterparts. When violence does occur, however, it is more likely to occur between males than between a male and a female. Our hominid ancestors would have observed, inflicted, and suffered these characteristics firsthand, but they would have also recognized them in many of their prey species, especially mammals. As a result, our learning biases with respect to the size/sex relationship and male-on-male violence in our own species would have been reinforced by our encounters with other species, particularly those upon which we most depended for our own survival. Still, the question remains: if males tend to be larger than females in my own species and I am thus predisposed to relate size to sex, why did I automatically assume that the jumping spiders, one of which was twice as big as the other, were both male?

My initial interpretation of the male's palpitations as a sign of aggression was incorrect, but my perception that they were aggressive precipitated a corresponding frame of mind. Once that happened and I was locked into my perception, it was very difficult to generate other possibilities. This may explain why even after the larger, female jumping spider appeared, I assumed she was a rival without giving it a second thought. But what else might have made this transference possible, considering the remarkable size difference between the two spiders, which should have alerted me to the possibility that I was seeing a male and a female? On the whole, like the males of other species, human males abide by certain protocols that make it unlikely, for instance, that a larger male will attack a much smaller one, or that a much smaller male will initiate conflict with a larger one. However, males of nonhuman species are generally much stricter in their adherence to these protocols because males that sustain serious injury during conflict stand a good chance of dying.

Although it is unusual for unevenly matched males of any species to engage in prolonged physical combat, it is more likely to occur between humans. The moment I decided the two spiders were males engaged in conflict, there was very little, apart from the professor, that could have changed my mind. Before anything else, it seems I am a creature of habit. But each encounter with my spider neighbors increases my potential for getting things straight. Just as there are costs for failing to realize our potential with respect to the nonhuman world, there are costs for succeeding, too, if only because the more we realize about other creatures, the more we tend to care for them, and the more, therefore, we stand to lose.

Our success depends on the species. Without webs to commit them to a place, jumping spiders are a roaming species, which makes it more difficult to find and observe individual spiders on a regular basis. Just last night I spotted a jumping spider in one of our tomato plants, which, although it has long ceased yielding edibles, provides a number of creatures—including a large praying mantis and several tiny white spiders—with homes and hunting grounds. Although my interaction with the jumping spider was brief, it was a pleasure to watch it jump from stem to leaf, investigating his or her territory.

My relationship with the black widow is another story. I've been watching her now for about three months, which is a long time in the life of a spider. I have watched her mature from a juvenile to a full-fledged adult. I haven't always been sure of her whereabouts, however, nor of her well-being. One night, I went outside to check on her and she was nowhere in sight. Normally well-tended, her web was in disrepair. Not a good sign. I flashed my light on the entrance to her lair and saw nothing. I anxiously wondered if she had been killed.

Thinking that she might be coming out after I had gone to bed, over the next three or four nights, I went out a few hours later to see if she had repaired her web. Instead of improving, the web fell into greater and greater disrepair, until finally all that was left was the scaffolding. I was convinced she was gone, but then thought back to the last night I saw her and realized there was another explanation. That evening she had done very well for herself, capturing a cricket and a large cockroach, as well as three honeybees that were, strangely, wandering about long after dark. I went out the next morning expecting to see her victims wrapped in silk, but they were gone. Not until a few days after the widow had disappeared did I remember this last detail. Once I did, however, everything began to fall into place. I remembered the appearance of her abdomen in the days leading to her disappearance, how it seemed inflated, like a tiny black balloon. Perhaps she was preparing to spin her egg sac, in which case her cache of prey would get her through the lean days ahead. There was only one way for me to find out.

According to my calendar, the last night I saw her was September 29. Four days later, on October 2, I decided to test my theory. I went out about an hour after dark. The evenings were cool, so as usual, the cats were eager to join me. Perhaps a bit too eager. We have two cats: Winston, an eight-year-old male, and Bella Jean, a one-and-a-half-year-old female. Winston and I have lived together for a long time and know each other's routines pretty well. He tends to give me plenty of space when I'm trying to conduct my research. Bella Jean, however, might chase the orb of the flashlight right into the widow's web. On the evening of October 2, however, there was no web that she could destroy.

I don't know why it hadn't occurred to me sooner, but on that night I decided to lie on my belly and flash my light into the widow's lair. A corridor stretched some ten inches straight back into the bush. At the end of it, suspended in the air like a tiny moon, was an egg sac. I didn't see the widow at first because she was behind the sac. Then I saw her slender legs cradling the precious contents. Today is October 13 and she is still guarding her brood. After two weeks of maternity, her once swollen abdomen is now not even half its previous size. She appears shrunken and lackluster, a living ruin.

I have not seen the widow for the last three days, nor have I seen her egg sac. I am not too alarmed by the disappearance of the egg sac, however, because she had on one occasion moved it to some other chamber in her lair, and a short time after promptly returned it to the original spot, where it remained until three nights ago. And so here I am again, wondering and worrying a little. As a result of my first experience with her disappearance, I must say that I am now more open to the possibility that life, although hidden from view, goes on somewhere deep inside her lair. Perhaps one of the challenges of trying to understand a behavior is not to become so focused that we forget that the behavior occurs in the context of the creature's entire life.

I began this inquiry by trying to understand why the black widow would appear at dusk even though doing so would seem to put her at risk. I speculated that she appeared when female crickets began their search for males. I went out night after night to time the appearance of the widow and the crickets. Over the next two months or so, my hypothesis was supported by the fact that the widow appeared roughly an hour before the crickets started wandering. Consequently, I developed expectations—so much so that when the widow failed to appear that night after gorging herself on honeybees, a cricket, and a cockroach, I panicked. One exception to the rule called into question the accuracy of two months' data.

The inability to anticipate the anomalous can be dangerous, but the danger is not solely academic: it is one thing to lose two months of data, but it is quite another to have one's life threatened. A few days ago I was outside on the patio rearranging some furniture and plants. Kim and I bought several plants to adorn the patio area, one of which is a purslane. Initially I hung this plant to encourage growth, but after the hanging pot broke and the plant fell, I placed the plant in a clay pot and set it on a small patio table. The plant was looking a little weak, so I moved it to the other side of the patio where it would enjoy more sunlight. Potted, the plant weighs about twenty pounds and requires two hands to transport it. The thick, leafy stems grew out over the sides of the pot, so I slid my hands between them and the pot to avoid damaging the plant. A couple seconds after doing just that, I felt something prick my right index finger.

At first I thought I had poked my finger on a piece of plant debris or on the pot itself. That explanation changed, however, when I noticed my finger was numb in the area where it had been pricked. I returned the plant to the table and carefully lifted the leafy stems. I didn't see anything, so I turned my attention to my finger, which was becoming increasingly numb. As I studied the endmost segment of my finger, I saw what looked like a tiny puncture. Turning back to the plant, I lifted the stems above the rim of the pot and saw the black widow.

She was tucked into a position that was both defensive and anticipatory, as though she were waiting to see what I was going to do now that she had bitten me. I couldn't help but see the humor in being bitten. Over the last couple of months I had not necessarily begun to think of myself as an expert in spider behavior, but I knew I was a more knowledgeable arachnologist than my in-laws, for instance. I also pretty much knew that I was going to be bitten. Roy Horn was mauled by a white tiger. Steve Irwin had been impaled by a manta sting ray. I was bitten by a black widow. The humor was short-lived, however. My amusement turned to fear, which turned into mild terror, which, as the moments passed, threatened to mushroom into a full-blown panic attack.

I had dealt with panic before. As I walked into the house to put on some clothes for my trip to the emergency room, I told myself to be cool. In the end I decided against going to the ER and instead knocked on my neighbors Iliya and Brenda's door.

I tried to hide my fear when Brenda opened the door, but from the look on her face, I wasn't doing a very good job. Brenda is a nurse, so she was careful not to show any alarm that might exacerbate my anxiousness. After agreeing to drive me to the ER if necessary, she gave me the number to the Center for Poison Control and advised me to call them for assistance.

I felt better knowing that Brenda knew I had been bitten, and I was grateful that she was there to help. I also felt a little annoyed, however. Although Brenda meant well, as she wrote down the number for the CPC, she reminded me that sometimes it takes hours for the venom of a black widow bite to take effect. I was just trying to deal with the crisis of the moment, and here she was reminding me of what might happen hours from now. Maybe I didn't need neighborly insta-care after all. When I got back to my house, I called Kim at work to let her know that her husband and the father of her soon-to-be-born child had just been bitten by a black widow. First I made the usual inquiries about her day and then I said "You're not going to believe this."

In reality, nothing could be further from the truth. She's come to expect at least one or two "unbelievable" stories from me each time I return from a hike in the desert or a fishing trip in the mountains. (Two days earlier, for instance, I was walking in the Superstition Mountains and was charged by a javelina. Kim was not thrilled.) She has always been interested in my well-being, but now that we are about to have our first child, she is especially interested. I tried to hide my nervousness as I explained what had happened.

When you've known someone as long as Kim and I have known each other—almost eighteen years—it is very difficult to conceal anything, even such seemingly private things as emotions. Once the voice gets involved, what is difficult to conceal becomes impossible to conceal. I wasn't surprised when the timbre of Kim's voice began to match my own. I explained what had happened and what I intended to do. Kim did a much better job of dealing with her worry, asking if I had any symptoms, and if I was sure it was a black widow. I was anxious to contact the CPC, so I did the best I could to quickly answer Kim's questions: apart from the numbness in my finger, no, I wasn't experiencing any symptoms. And yes, I was sure it was a black widow.

Kim insisted that she stay on the line while I called the CPC. I called and talked to a man whose name I have since forgotten. Once he had acquired all the necessary information, including my age, health, and when and where on my body I was bitten, he told me symptoms to watch for and to go to the emergency room should I experience any of the more serious reactions—including vomiting, chest pain, and respiratory difficulties. He made a point of reminding me that these latter symptoms usually only threaten the very young or the very old. I was relieved—until he said he'd call me again in a couple of hours to make sure I was alright.

Over the next two hours I did what I could to relax. I started thinking through what had happened. I was encouraged by the fact that the spider was still a juvenile. In contrast to the adult females, the juveniles are typically brown with four crisp, boomerang-shaped slashes of white on the top of their abdomens, but I have seen other juveniles with two parallel white slashes flanked by a red slash on either side. Their beauty is strikingly reminiscent of the etchings on Navajo pottery. Apparently, the juveniles of both sexes are harmless. Males don't have venom, and even if they did, they are so small that their fangs could not even penetrate the skin. Females do have venom, but it's only when the female reaches adulthood and turns black that she becomes really dangerous.

This knowledge helped a little, but then I started wondering about the line between the juvenile and adult stages of the female widow's life. How fine is it, and at what point does her venom become dangerous? The doorbell rang before I could pursue these thoughts. Iliya, Brenda's husband, had stopped by to check on me. I told him that, apart from a mild numbness at the site of the bite and a little anxiousness, I was fine. I joked that perhaps I had some immunity, that I would become the marvel of the arachnological community. "That's good, man," he said in his soft-spoken way, "but remember it sometimes takes hours for the venom to take effect." I fantasized about slamming the door in Iliya's face, but I just smiled and said, "That's what I hear."

After Iliya left, I returned to my ruminations. Why was the black widow in the plant in the first place? I have seen twenty black widow webs around my house over the last two years, and not one of them was constructed in a place like the purslane plant. Actually, there wasn't really a web at all in that plant. No scaffolding, no capture fibers, just a juvenile black widow and a scraggy rag of silk no bigger than a quarter. The data had simply not prepared me for this anomaly. But that is the nature of data. For even while data are used to gain insights into the nature of the species as a whole, the whole itself is comprised of individuals, any one of which may for whatever reason deviate from the average measure of the population.

I was lucky: for a number of reasons I will never know for sure, apart from a mild numbness at the site, I did not have a reaction to the black widow's venom. This may have been because she didn't have the venom to inject in the first place, or if she did, realizing I was not prey, she injected only enough to warn me. This seems unlikely, however, when I consider the relative lack of incentive other juvenile poisonous creatures (e.g., the rattlesnake) have to regulate their venom. Perhaps the adult would control the amount of venom injected, but the young, which are not large enough to be intimidating and lack the experience needed to distinguish threats, and which are therefore more vulnerable to predation, may not be nearly as discriminating. The other possibility is that I have some resistance to the venom. But I wouldn't bet on it.

A few days later I went to check on my assailant and discovered she was no longer there. Surprisingly, I was disappointed by her absence. I was looking forward to watching her molt into a full-fledged adult, just as I had seen happen to the black widow on the east side of the house. I admit that after being bitten I considered destroying all the widows around my house, but thinking it through, I decided doing so would somehow violate my contract with the different species of spider in my yard. And of course it wasn't the spider's fault. Black widows, the jumping spiders, the marbled spiders, and the little orb weaver with her web and its stabilimentum—they all were important to me. Through my observations of the orb weaver, I gained an appreciation of all spiders and I wasn't about to sacrifice all this in response to a mini-envenomation. Though it may seem human life and the lives of other creatures are separate at worst and contiguous at best, my backyard adventures suggest that the extent to which our lives overlap and intersect is far greater than any distance between them.

The central question, then, is how do we realize this potential? It helps to keep in mind that our reactions to other creatures are both biased and mediated by our biological past. That I struggle to notice other creatures illustrates the depth of the difficulty we face as a species. But as I think my reflections show, it is possible to recognize and compensate for our predispositions. Perhaps that is one of the great gifts of ecological thought: as the study of relatedness, ecology enables and encourages us to notice specific aspects of the world around us, including other creatures. Rather than thinking of other creatures as isolated, ecology posits them as intimately tied to their environment. One could say that individual creatures are expressions or embodiments of their environment, so much so that if one wishes to know more about the creature, one must also understand the environment on which the creature depends.

Our biology is not the only barrier to our enjoyment, however. Cultural attitudes represent a completely new set of challenges. Here in North America, for instance, we tend to classify most animals as either cuddly, Bambi-like creatures, or as incarnations of evil with harmful intentions. The other animals of the world (and, don't forget, tornados) would appear to fall somewhere between these two extremes—that is, assuming we acknowledge them at all. Interesting how irrelevant these classifications are to the spiders I encountered in my yard. They don't even come close to clarifying the lives of these creatures. Despite our attitudes here in this country, however, even the most cursory look at cultural attitudes toward the spider shows that human culture need not be hostile or indifferent toward, nor misrepresent, other creatures.

My neighbor Iliya, for example, is from Nigeria. His people believe that spiders have great wisdom, and they are therefore treated with great respect. But I do not need to look as far as West Africa to find examples of cultures that revere spiders. Here in the Southwest, the Navajo people revere Spider Woman, a god-like being that created all other life forms, including humans, and then connected them with threads of silk. This is a beautiful story, to say the least, but it is also humbling: in a remarkable reversal of a popular belief, Spider Woman, not a white haired, white robed deity, created humankind. Then she placed us together with all the other creatures in the web of life. According to this a-hierarchal view of life, then, we are special—but so too is everything else.

These and other stories show that our accounts of other creatures can teach respect without engendering fear and hostility. But, as I think these accounts and my own bourgeoning, ecological perspective show, the degree to which we respect other creatures and ourselves, ultimately, is deeply informed by how well we understand the world we share. True, understanding another creature is extremely difficult (I hear the astronomers exclaiming, "You've got it so easy!"), but it is wonderful that we can pursue such an all-important and rewarding ambition in our own backyards.

A FIELD GUIDE TO HABITAT THEORY

i. The Pine Grove Saga

I lived in Caribou, Maine from 1968 to 1979. The town was so small it had only two movie houses. The first was a single-screen theater a few blocks away from my home on Sweden Street. Caribou Cinema, by contrast, was on the other side of town, so my siblings and I rarely went there. Of the two theaters, Caribou Cinema was the nicest: it had two screens, popcorn machines, bright red carpet, and colorful posters that advertised current and coming attractions. I remember standing in front of one such poster with my brother and sister until my mother hurried us into the theater. The poster advertised a movie called *The Bermuda Triangle* or something on that order, but I remember with certainty that it depicted a woman from her midriff down. She was wearing a yellow bikini bottom (hence the triangle), and a tiny ship with enormous sails was about to sink beneath her bikini line.

Adult or otherwise, movies were a big deal when they came to Caribou. They were messages from the outer-world, or at least other worlds, and offered reprieves from life as I knew it. Nature horror films (films in which nature was depicted as the villain) were a favorite because they added a new element of wonder to country life. Then life would get back to normal until the next movie came out. I knew not to invest too much in films that portrayed nature as the enemy of man (women are not generally known for their abuse of the land). As far as I could tell, both from the films themselves and from my own experience, man—because of his often violent and brutal ways—is the greatest enemy of nature. I think this is why, of all the films that came out during my childhood in Caribou, *Jeremiah Johnson* (1972) made the greatest impression on me. Although I wouldn't have been able to say so at the time, what attracted me to the film—and why I remember it so well even now—was its depiction of Johnson's relationship with the natural world and with Native Americans. Unlike the horror films from the early seventies, *Jeremiah Johnson* did not demonize nature. Instead, Johnson's character underwent a process whereby he learned, finally, to co-exist with the natural world and with other people (largely represented by Native Americans) on their own terms.

I don't think this difference between horror films and *Jeremiah Johnson*, a drama, can be explained away by genre differences, either. While the former are not characterized by their truthfulness, the fact that the majority of horror films features at least one enlightened character, usually a scientist or Native American, who understands the natural phenomenon or "monster" with which the tiny humans must contend, suggests that film-makers have some options when telling their stories. That no one ever really listens to the scientist's warnings is not merely a filmic device for creating tension, or a reflection of cynicism toward scientific knowledge; it is a reflection of the world's failure to acknowledge our mistreatment of the Earth.

Instead, the characters defeat the villain, whether it is a shark, a swarm of bees, an anaconda, a volcano, or, as was recently depicted in the movie *The Grey*, a pack of bloodthirsty wolves. The main characters experience no significant change apart from a few cuts and bruises. Maybe someone falls in love. But They gain no increased awareness of the nonhuman world, represented by the monster, or of their impact on it. I haven't seen any horror films lately, but from what I can tell, there seems to be more effort to present humans not as isolated and special super-creatures, but as a species that has the unbecoming habit of fouling its own nest.

I was not thinking about any of this as a child. I was much too busy being awed by *Jeremiah Johnson's* depiction of the American wilderness and the native people who inhabited it. Although I am now well aware of the erroneous and damaging effects the concept of the noble savage had and continues to have on Native Americans and environmentalists, as a child I only knew what I saw, and I was mesmerized. One scene would come to inspire many days of play. Toward the end of the film, Johnson is fishing a beaver pond when a Crow warrior approaches him. Although the scene was memorable for several reasons—including the golden grass and the black water, the implicit smell of thawing earth, the sun's warmth on Johnson's clutched hands, and how I dreaded the going of the late spring light—the Crow warrior seemed borne of land, sky, and water. He wore tawny skins and raptor feathers. Black and white war paint transformed his face and eyes into alpine lakes fringed with snow. His horse moved like an apparition. Together, horse and rider seemed like the commingling of wilderness and human desire. I distinctly recall hoping that this Indian would prove friendly as he rounded the pond, dismounted, and gazed over the back of his horse at Johnson. But Johnson was an Indian killer. Predictably, a few moments later, the beautiful Crow warrior was dead.

Although Johnson was at times brutal, I became very attached to him, the Crow warrior, and the wilderness, which was as important as any human character. If anything, the wilderness co-starred in the film, because Johnson could not have become who he became without it. Thus, Johnson was an expression of the place. But Johnson is not alone in this; people the world over are where they live. As a child growing up in a small farming community in northern Maine, however, as long as there were woods and places where I could live out my imagination, I would happily trade in my work boots and overalls for the Crow warrior's skins and feathers. I wanted to live in that world so badly that when the film ended, I rallied the support of my brother Christian and our friend Tom, and together we found ways to sustain the world of *Jeremiah Johnson*.

Living in a small country town, we learned the importance of entertaining ourselves. We quickly developed a game enabling us to continue the Johnson legacy. Before we could play, however, we had to sort out some logistical problems. We needed at least three people—we had that with Christian, Tom, and me. Deciding when to play was easy: anytime, so long as we had enough time to play until the game ran its course and we were home by dark. Given the nature of the game, we needed privacy. For this reason we chose to play in a grove of conifers someone had planted out past the potato fields near Tom's house. (Actually, to this day I'm not sure if the grove wasn't actually a tree farm. I would like to believe that someone just really liked trees.) There must have been fifty rows of trees with one hundred trees per row. Stepping into grove was like stepping into dusk in the middle of the day.

While the right location was very important to making the game seem real, we did more than select a place. Not long after my brother and I had seen the film and began fancying ourselves mountain men, my father bought us each a mountain man jacket. Thankfully, once we had outgrown them, my mother made a point of properly storing and safeguarding them. Nearly thirty years later, I have mine hanging on my office wall, which is where it will stay until the day my son Wilder needs it for his own adventures. The jacket is made from cowhide and—apart from fringe across the shoulders, chest, and pockets, and a satin liner—is simply designed. The left sleeve still bears a vague, white stain from one of the many wet winter days of my childhood. This little jacket served me well, but it also presented me with a problem.

Unlike Tom, who didn't own a mountain man jacket and therefore had to be an Indian, each time we played the game,

I had to make a decision. What I find so interesting is that we would even want to be Indians, especially since many of them were portrayed in *Jeremiah Johnson* and other films as indiscriminate killers of white men, women, and children. In fact, most films were not overly concerned with presenting the Native American perspective, the consequences of which are still apparent today. Killers or not, in my child's eye it didn't matter. They were beautiful in their skins and feathers, and their moccasins seemed almost magical in their ability to hide their footfalls and protect their feet from the icy streams and mountain snows. Like all those who have hunted or been hunted throughout time, the Indians moved silently, a skill they shared with other animals of the wilderness.

Compared to the Native Americans, Johnson, while possessing his fair share of natural prowess, seemed to view the natural world as something to be subdued and controlled, a tendency that limited his relationship with it. Although in reality the Indians were not more or less natural than Johnson, they were shown as being in tune with the natural world, which they ostensibly accomplished by taking only what they needed to survive. Thus, I wanted to be the mountain man, the Crow warrior, and the animals whose skins they wore. I finally resolved the conflict by wearing the jacket until we reached the grove. Once there, the three of us would undress on the tree line, hide our clothes beneath the brush, and don our Indian garb of bootlaces and washcloths. Then we would step out of the sunlight and into the trees and the tone of our lives would change.

Once we had the proper place to play, we simulated Johnson's trials and tribulations with a mixture of tag, hide-and-seek, and what would later burgeon into a game of its own—stick fighting. Tom and I were a year younger than my brother and roughly the same size, but my brother was the strongest and the fastest. We had to find ways to offset his advantage. I was pretty quick, and sometimes I could shake off my pursuer with a series of quick turns and bursts of speed. But that defense was exhausting, and so while on a few occasions I managed to elude even my brother, usually he would chase me until I tired. Then I would have to face the consequences. My brother was typically gentle and did not bully me, so I wasn't afraid of being caught—not physically afraid, anyway. That is, my brother did not have to punch me in the arm or throw me to the ground and force me to eat dirt: he knew that in the complex world of boyhood, being caught was punishment enough. It was humiliating to be caught, but we knew instinctively that, had we been captured by a real enemy, things would not have ended well for us.

Luckily for Tom and me, victory was still possible. Nine times out of ten, these victories involved sticks. We all understood that the game would not begin until each of us had found a suitable stick. After undressing and donning our bootlaces and washclothes, we searched for weapons. In theory, these "swords" (in addition to being a deviation from the *Jeremiah Johnson* script) were to be used for hand-to-hand combat. When it came to actually fighting with them, however, the thought of being hit in the hand usually lead to a prompt retreat by whoever had the inferior stick, or whoever was less aggressive. Sometimes Christian could get away with having an unimpressive weapon due to his strength—the key to subduing him was to stop any hand-to-hand combat before it started. If Tom and I couldn't outrun him or best him through superior sword-play, we could still throw things at him and drive him off. We threw anything that wasn't rooted to the ground, but pine cones—especially those heavy with seeds—and short sticks were favorites. Rocks were off limits. Throwing sticks and pine cones may have been dangerous and stupid, but compared to the danger of being struck by a rock (my first rock fight was my last), throwing wood seemed almost reasonable.

Given the dynamics of the game, including the need for

privacy, it makes sense that we selected this place. Still, there were other factors involved in our decision-making. We were, after all, trying to recreate a world through our play—a wilderness filled with enemies, unknown territory and dangerous animals. Because different environments—the ocean, a forest, a mountain—inspire different emotional responses, wherever we decided to play, the place had to make us feel like we were living in the time of *Jeremiah Johnson*. For it would have been hard to believe we were running through the woods of the nineteenth-century American West if we could hear cars or Tom's mother calling him home for an early supper. But the knife of privacy cuts both ways: if we were isolated from adults and any interruption they might impose, we were also out of reach of help should something truly dangerous happen. Children are not normally burdened with excessive worrying, so this concern was more latent than anything. Closer to the surface, however, were other feelings that made the pine both attractive and repellent.

Ten yards into the grove, the light level fell to near-darkness. Like the grove's isolation, the poor light inspired ambivalence. During those times when the game would shift into hide-and-seek mode, darkness proved very useful. But the moment I left my hiding place and was again on the move, the darkness hindered me as much as it helped. Moreover, the canopy of the closely seeded trees created a shady reprieve from the summer sun. If there were any real enemies or dangerous animals in the area, however, it followed that they might also be in the grove seeking relief from the sun. I would no doubt consider these trees a place where I might find prey if I were hunter with a lethal weapon. Great white hunter I was not: I was just a little boy with nothing but a handful of pine cones, a stick, two washcloths, and bootlace between me and anything I might encounter. When I was in those trees, I felt as much like prey as I did predator.

The trees were planted roughly ten feet apart, and the rows

were separated by about twelve feet. This spacing resulted in long, north/south corridors that stretched out of sight. Notwithstanding the discomfort of running barefoot over conifer needles, the corridors were relatively easy to travel, and most of the chasing and running took place along them. The quantity of these corridors, coupled with the vastness of the grove, would suggest openness, but navigation was actually very limited. The trees had grown branches of decreasing size from the base of the tree up, and due to insufficient spacing, they tended to grow into one another, forming a nearly impassable mesh of sharp branches.

If I couldn't run through the trees, perhaps I could climb them, high above my brother's reach, and well beyond his ability to follow. Unfortunately, because of the density and size of their branches, these trees did not lend themselves to climbing. The only unobstructed way for me to escape was to run down one of the corridors and hopefully have time to fight my way through the pine branches into an adjacent corridor (running out of the grove wasn't an option: if one of us did so, we automatically lost the game and, as it happens, the respect of our playmates). More often than not, if I were being pursued by my brother, taking the corridor usually ended with me being outrun and, finally, caught. In the end, the grove provided too few opportunities for escape.

To make matters worse, once we were inside the grove we could not see the sky. We were not only physically confined while in the grove, we were mentally and symbolically confined as well. Without being able to see the horizon, or the proverbial light at the end of the corridor, it was hard to imagine a way to escape the threatening world of our game. Perhaps that is the nature of claustrophobia: the fear intensifies when both the body and the imagination cannot escape, a premise that is surely built into edifices such as shopping malls, as well as many of our maximum security prisons. Interestingly, the grove had a kind of prison cell equivalent. At the very center of the grove was a small clearing, a quarter acre in size, where we would often meet to eat our snacks and warm up. The clearing had its appeal, but the solace I felt there was short-lived: being in the clearing was like sitting on a raft in the middle of the ocean—comforting, as long as I didn't think too much about the forest surrounding it.

However much this description of the grove suggests otherwise, we didn't play there very often. It seems the place's repellent features proved stronger than its attractants, even though many of our apprehensions—such as the presence of dangerous animals and real enemies—were unfounded. As a whole, the grove was relatively featureless, which makes me all the more curious about why I remember it so well. One explanation may lie in the apprehensions themselves—we tend to remember things that make us feel uncomfortable and afraid. Certainly my apprehensions about the place affect my memory of it. Odd as it may sound, however, I also think I remember the grove so well *because* of its featurelessness.

I admit this seems like a paradox, for how does one remember what appears to be so unmemorable? Perhaps when featurelessness is the cause for apprehension. This explanation seems especially plausible given that the grove was, for us, a novel environment. Had we played in the woods near Tom's house, he would have had a slight advantage over me and Christian, and the same would have been true for Tom had he walked down the hill and played in the woods below our house. In the grove, no one had the advantage that comes with knowing a place, including the most efficient travel and escape routes, which is one reason why no one suggested we regularly play there.

Apart from the clearing where we ate our snacks, the grove had no distinguishing features or landmarks that we could use to help us orient ourselves or navigate the terrain. The grove was an indistinct environment. Whether we use landmarks to assist us in finding food, escape danger, or find shelter, selecting and remembering distinctive landscape features is crucial to fulfilling our basic needs and, therefore, to our survival. After our last walk in the Superstition Mountains, my mother would agree. She lives in Salt Lake City, where, in addition to understanding the logic of the city's layout, she has grown accustomed to having the Wasatch Mountains as points of reference. I have lived in Salt Lake for many years myself, and as long as I can see the mountains, it is impossible to get lost. Having never stepped foot in the Sonoran Desert, my mother was out of her element. I knew she was nervous after the third time (within an hour!) she asked me if I were sure I knew how to find my way back to the car. "Yes, mother," I reassured her. "The car is just over that hill."

Considering my mother would have seen houses had she looked east, it's tempting to conclude that her feelings of vulnerability were just plain silly or harbingers of an aging woman's dotage. Even assuming I didn't have a cell phone with me (which I did), couldn't we just walk to one of the nearby houses and ask the occupants for assistance? And what about the apprehensions I felt while playing in the grove? Should they not also be dismissed as irrational kid fears? Perhaps my mother and I are unique in our responses. But is she the only person to have felt the profound sense of anxiousness that attends the thought of being lost? Was I the only child to have felt afraid in a dark place, or to have needed a way to escape the creatures of my imagination? Perhaps we describe as irrational any response for which a cause is not immediately apparent, but that does not mean the response is unfounded.

Actually, although the majority of environmental dangers have been neutralized or eradicated, people continue to respond to certain environments and situations as though the dangers were still present. In his essay "Biophilia, Biophobia, and Natural Landscapes," Roger S. Ulrich notes that "findings from many laboratory conditioning experiments support the notion that humans are biologically prepared to acquire and especially to not 'forget' adaptive biophobic (fear/avoidance) responses to certain natural stimuli and situations that presumably have presented survival-related risks throughout evolution." According to this view, then, the fact that my mother wasn't really in any danger of being lost is beside the point: what mattered is that she had that perception.

This is the subject of habitat theory, which Jay Appleton introduced in his 1975 book *The Experience of Landscape*. The theory suggests that our aesthetic preferences for, or aversions toward certain environments stem from the need throughout human history to select habitats according to how well they help us meet our biological needs. Appleton breaks the theory into three key concepts: prospect, refuge, and hazard. Basically, what makes an environment habitable depends on its advantageousness: habitat must offer sufficient prospects in the form of resources, as well as provide a place of refuge from what Gordon Orians calls "ecological dangers," including "predators, conspecific individuals, and inanimate hazards (storms and geomorphological conditions)."

Thus it is not just the perception of danger that links us to earlier periods of human history. For, as Ulrich's title and my own experience suggest, the converse of biophobia is biophilia, which was first described by E.O. Wilson as the "innate tendency to focus on life and life-like processes." If we are sensitive to those aspects of the environment that might harm us, it follows that we must also be sensitive to aspects of the environment that aid in our survival. No one could reasonably argue against the universal (and well-documented) human tendency to look for "fitness" cues when selecting a mate. Similarly, considering the intimate relationship humans have with the environment, it does not seem far-fetched to suggest we also evolved an ability to focus on and interpret environmental cues in order to best utilize environments.

If there is any truth to habitat theory (and I believe that there is) it would mean that the tendency to focus on prospect, refuge and hazard when deciding where to live would persist in modern day humans. In his essay "An Evolutionary Perspective on Aesthetics" (which I quoted earlier), Orians explains the tendency this way: "If one accepts that life on Earth has been molded by the long-term action of natural selection, it follows that the strong emotional response of organisms evolved because they, on average, increased fitness; that is, they improved the survival and reproductive success of the individuals that expressed and acted on them." But therein lies the problem.

Orians' provisional formulation hints at the difficulty we have in comprehending ("accepting" is further down the road) life's deep history, and that how we decide where to live might actually have something to do with the fact that humans have been deciding where to live for hundreds of thousands of years. But this could just be a matter of broadening our concept of genetic transmission. Kim and I take great pleasure in looking at our son Wilder and finding in his face traces of ourselves or of our parents. My mother has gone even further by tracing her own genealogy two hundred years or so, back to the Isle of Man. But if it is possible to go that far back in time, is it not conceivable that we could keep searching until we reach the threshold of human and pre-human history? Does our genetic lineage end where the paper trail ends? If not, then hasn't genetic transmission occurred for hundreds of thousands (as opposed to merely hundreds) of years? Perhaps, but even if this genetic heritage is conceded, we aren't like early humans. We have changed.

We aren't those hunched-over, brutish knuckle-draggers, those murderous monkey look-alikes in Stanley Kubrick's *2001: A Space Odyssey*, which, despite their newfound weaponry of femur bones, still cower in their caves, terrified by night's noises. We aren't even like the more sophisticated Flintstones in their leopard skins, eating fried pterodactyl eggs and brontosaurus burgers. Bedrock be damned, we are *Homo sapi-*

ens, man the wise. Whatever we are, it is hard for us to believe we are somehow related to all other humans (not to mention other animals such as the chimpanzee, a primate whose DNA is 98% identical to our own), especially if they lived hundreds of thousands of years ago. Considering that each one of us carries a set of about 30,000 genes, some of which carry information that is known to influence the development of eye color, height, blood type, color vision and much more, isn't it possible that the information encoded on some of our other genes "encourages" us to develop a nervous system that comes with environmental biases that have a survival benefit? If so, it would be worthwhile to investigate the contents of our genetic packaging to the fullest extent possible.

ii. Home Advantage

I have tried to demonstrate how biologically prepared biases may have informed my response to the pine grove where I played as a child and, to a much lesser extent, my mother's simultaneous delight and distress while on our jaunt in the desert. In an effort to further test habitat theory, I have been conducting informal research in the Superstition Mountains. After logging over 100 hours hiking in the desert, I was delighted to see habitat theory at work, both in my own responses and in the responses of my "subjects." This did not surprise me. The Sonoran Desert has many savanna-like characteristics, including views unbroken by dense vegetation and trees that bifurcate close to the ground and have broad canopies. What did surprise me, however, is how much habitat theory influenced the way I think about my home, and why we might interact with our habitat in the ways that we do.

E.O. Wilson once commented on how people will, when asked to rank landscapes according to their appeal, choose first the landscape that most closely resembles the place of their upbringing. Most people chose savanna as their second choice. We feel a strong attraction to the landscape features of humanity's first home, so the argument goes—the savannas of Africa.

In addition to resembling savanna, the Superstition Mountains have a long history of human inhabitance—another important point from the perspective of habitat theory. The early Desert Archaic peoples inhabited the area around 9000 BP, and later the Hohokam (a Pima word for "the vanished ones") did as well. Although the Desert Archaic peoples left the area more than a millennium ago, it is not unusual to come across places where they spent their time. One of the more remarkable places is a rocky knoll just off the Gold Canyon Trail, which is where I went when I first started wondering how habitat theory might help me clarify my connection to the environment. I could not have chosen a better site to begin my inquiry. As a place that offers prospect and refuge, the site fulfills the theory's predictions.

Perhaps the best way to understand the layout and appeal of the site is to imagine a multi-tiered, rolling mound of rock that rises above the surrounding desert. Directly east of the knoll is a large wash lined by an array of plants, bushes, cacti, and trees. The wash originates high in the Superstition Mountains, which rise just north of the knoll, perhaps a quarter mile away. West of the knoll, the desert—not quite flatland, not quite mountain—dips and rolls beneath a mesa. Mountains extend in all directions, but the mountains to the southwest and west—the San Tans, McDowells, Userys, White Tanks, and Maricopas—are farther away and are often concealed by a desert haze of dust and pollutants. On clear days, however, I can see out across the valley floor, which, except for the occasional island of rock, spreads flat and unbroken. This description would have been especially true for the Desert Archaic peoples and their successors, for apart from seeing their own dwellings, the view of the desert floor and the horizon beyond would have been unimpeded by the rooftops, power lines, and concrete now needed to accommodate millions of modern inhabitants.

One of these rooftops is attached to my home in Gilbert. Although I cannot see the Superstitions unless I stand on my roof, when I have stood out there—either at night to gaze at the moon and stars, or during the day—I tried to imagine life before we were here. Surrounded by the hustle and bustle of human activity, it is difficult to appreciate the vastness and the stillness people would have confronted before this valley was settled. But that is exactly what people would have experienced sitting atop this knoll on a clear day 9000 years ago. Much has changed since then, but whenever I sit atop these rocks and look to the east, south, and west, I realize that one thing hasn't changed: the deep-seated appeal of this spot and others like it.

I first learned of the Gold Canyon site a couple of years ago from my friend and former colleague Bert Bender. As the two of us sat atop the knoll eating our lunch and admiring the view, Bert placed his outstretched hand on the rock and asked me why I thought we are attracted to boulders and to other large rock formations. I suggested that perhaps our attraction to rocks had to do with their ability to provide shelter for our early ancestors. I was thinking in terms of our human ancestors, but Bert wondered if it had more to do with pre-human events. "I don't know," he said pensively. "What about how fish gather around a bed of seaweed or a piece of driftwood floating in the ocean?"

In terms of timeframe, Bert's speculation was more exciting to contemplate. We agreed that the attraction had something to do with shelter or refuge, but Bert got me thinking about some other factors that might be involved in our relationship with rocks and, consequently, habitat selection. When describing the Gold Canyon site, I initially used the word "knoll," but "island" might be more descriptive—few things in life are more terrifying than being adrift in the ocean.

This feeling of deep vulnerability seems to have been on the

Pleistocene mind of the director of *The Perfect Storm*: toward the end of the movie, Mark Wahlberg's character is swept off the boat. The scene is shot in such a way that the camera remains stationary in the massive waves. A moment later, the wave that carries Wahlberg's character comes from behind us and we see his face before he continues on, far into the vast expanse of rolling black ocean. Evoking a mixture of serenity, resignation, dread, and, oddly, an iota of hope, the image is sublime.

Deep Water, a low budget, yet highly effective film that came out in 2006, also tapped into this primal terror, except in this case the perspectives of the audience and the protagonists were one-and-the-same. That is, we were in the water with the characters, two divers that had to tread shark-infested water after they were forgotten and abandoned by a commercial diving outfit. I was so desperate for some security in the midst of all that peril that, apart from wanting to run out of the theater, I wanted the couple to dive down twenty-five feet to the reef below them for safety. The idea was absurd, but it illustrates the power of our need to shelter ourselves from vastness.

The security of any refuge—a coral reef, an island, a rock formation, my home in Gilbert—would be undermined if predators or enemies could easily find and access it. In other words, find my refuge, find me. However, a predator would have to be very eager to risk confronting an inhabitant on his own turf, which is where he is the most secure and, paradoxically, has the most to lose. This explains why few events capture our interest like news reports of home invasions. Sports teams also tap into this sense of power whenever they play a rival team on their home field.

Whether we stand to lose our lives or our team ranking, having a lot to lose normally translates into having a lot to fight for. Such determination in an intruder is frightening because this primal deterrent is ignored. Kim and I have talked about what we'd do in the event of an intruder, and it was during that conversation that I realized I had already implemented several security measures, albeit unconsciously. Our sleeping arrangement is illustrative: In every bed Kim and I have shared—and there have been at least a dozen—I have automatically slept nearest to the bedroom door. There are several explanations for this, including the intense need I feel to protect my mate and mother of my offspring, which I accomplish by placing myself between her and anything that might walk through that door. The hope is that an intruder would never reach this point, that our own built-in, "early warning" auditory system would have done its job and alerted us to any danger long *before* it reached us. How this detection works in my own case is curious.

iii. Assume the Position

I enjoy taking the occasional daytime nap or siesta. I've noticed my sleeping position varies, but typically I will lie on my left side and face the north window. I'm not nearly as flexible, however, when I sleep at night, when I must lie on my right side if I expect to get any shut-eye. What could possibly be going on here? Before I offer any speculation, it will help to know a couple more facts: first, when I lie on my right side my best ear (my left) is exposed and, second, I face the direction of the door. This orientation enables me to detect abnormal sounds within the house and to detect any potential threat more quickly.

If it weren't for my bad ear, perhaps I would—like the people in mattress commercials or who lie in hospital beds—be more inclined to sleep on my back. This would seem to have greater defensive potential because both ears are exposed and I could respond more rapidly and effectively than if I were lying on my side or my stomach. Sleeping on one's back is by most accounts better for the spine, yet I don't think I've ever known anyone who habitually sleeps on his or her back, and the majority of people depicted on television and in film also sleeps on their stomachs or sides. So we seem to have a bit of a puzzle here. Given the apparent benefits to the human spine, we would seem predisposed to sleeping on our backs, and yet most people I know say this makes them feel "exposed." And I agree. By exposing the throat and organs, the position is one of great vulnerability. This suggests that, for many of us, a defensive sleeping position trumps one that might allow a quicker transition to an upright position. But that doesn't sound right—how could the position be truly defensive if it undermines readiness?

I think most people would agree with the idea that we are most vulnerable when we sleep. Ideally, a defensive sleeping position would be tailored to the worst-case scenario. Assuming a predator managed to stalk its way to our bedside and attack us, the likelihood of a fatal bite or blow might increase if we were on our backs, a position that exposes the soft flesh of the throat. Perhaps the difference seems negligible, but to the animal that, by sleeping on its stomach or its side, increases its chances of surviving an attack, the difference may be between life and death.

My wife Kim has significant training in biology and was learning about natural selection at a time when I was dashing off poems, totally absorbed in American, British, and German Romanticism. On one of our daily walks with Wilder, I shared my sleeping position hypothesis with her. Although she did not quash my speculations, she did help identify their limitations.

Kim asked me if I had considered our quadrupedal ancestry. By her estimation, our tendency to sleep on our sides and stomachs may be a remnant from our pre-human past. I thought of all the different animals I have seen in my life, and realized I could not recall a primate, mammal, ungulate, bird, or reptile that habitually and continually lies on its back when sleeping. I know there is at least one species of snake (the western hognose) that, as a defensive strategy, feigns death by flipping partially on its back and emitting a noxious odor that mimics the stink of decomposing flesh. Also, wolves and other canids will lie on their backs to indicate submission. Whether it is dead or living, sleeping or waking, sick or healthy, the one thing an animal communicates when on its back is vulnerability.

We are at the height of our vulnerability when sleeping (a close second is when we are mating), so, like other animals, we would expect to sleep in a position that increases our chances of survival. It doesn't follow, then, that by lying on our sides or our stomachs we are undermining our readiness. We would not have adopted a sleeping position that makes us more vulnerable. Adaptive behavior should increase fitness, not undermine it. Kim's quadruped hypothesis got me thinking about the different ways of lying down and then returning to an upright position, and how those differences might affect our ability to respond to danger.

So I conducted a little experiment. I tried each position: I lay on my back, side, and stomach. Then I rose suddenly—as if I had detected some threat or danger. In terms of how they affected my ability to physically *engage* an attacker, there was very little difference between the positions. The difference seems a bit greater, however, when one is lying stomach-down. It would be difficult for me to address a danger if my back were turned. So how might natural selection have evened the odds? If sleeping on one's side or stomach is truly adaptive, these positions should increase our chances of surviving an array of attack scenarios.

Thinking in terms of prevention (as opposed to direct engagement), the difference between lying on one's back and lying on one's side or stomach becomes substantial. Most animals, when confronted by danger, will instinctively utilize the flight response. We are no exception. Our goal is to put distance between ourselves and whatever is threatening us, be it a storm, predator, or another human being. By doing so we buy time to assess the threat and decide what action, if any, is to be taken. Most animals share the goal of avoiding dangerous situations. Therefore, we can predict that animals prefer resting positions that offer the greatest likelihood of escape. Although humans are bipedal, our bodies are still functionally similar to our quadrupedal ancestors. I do sit-ups, but it is still considerably more difficult for me to rise from my back to a position of readiness than it is to rise from my side or stomach, which is when we have the greatest cooperation between major muscle groups. This would be especially true if I were rising from the ground, which is where we slept long before the advent of the mattress.

If these speculations are at all plausible, then we may tend to sleep on our sides or on our stomachs because those positions facilitate mobility and escape. In the event of surprise, the flight instinct remains, but it is usually shortened to the time it takes to turn and engage the danger. I can't help but wonder what other commonplace phenomena might be better understood by asking how they are adaptive. Because a defenseless animal is soon a dead animal, we would expect to have developed other ways of maximizing our ability to detect and respond to threats and thereby secure ourselves in our environment.

iv. Singing in the Morning, Singing in the Evening...

Feeling secure in a place involves knowing its sounds. Each time I have moved into a new dwelling, I don't think I slept much for the first two weeks. When we first moved to Gilbert, every noise was novel and reason for (largely irrational) concern. Now that I know what to expect, I tend to hear only anomalous sounds—the rest are relegated to the sound heap that is white noise. Of course, not all anomalous sounds translate into threats but we still notice them. For instance, for the last five days a mockingbird has perched in our orange tree and has sung in earnest from about 2:00 to 4:00 in the morning. I like hearing mockingbirds sing, but not at 2:00 a.m.! Still, after about a week of hearing this nocturnal concert, I can more or less tune it out. Without this ability to filter sounds on the basis of their importance (for instance, the sound of screeching tires compared to the chirp of crickets), I don't think any of us would get more than a few minutes of sleep at a time.

Whether we wake or sleep, it is natural to prioritize our sensory experiences. For me, this means paying close attention to the gate on the west side of my house. The west side of the house is also the west wall of our bedroom and master bathroom. The wall features two rectangular windows and one larger window above the bathtub. The north wall of our bedroom has an enormous window without a sunscreen. Despite their apparent vulnerability, however, these windows are not the primary security concern, which instead is the gate to our backyard. In terms of prioritization, it makes little sense to focus on our bedroom windows when an intruder must first go through the gate to reach them. Any point of entry represents vulnerability in our refuge—we've had to find ways to minimize and deemphasize these vulnerabilities.

Some obvious examples include security systems (electronic and canine), and locks on our windows and doors. Not so obvious may be our tendencies to sleep in a certain position, on a particular side of the bed, and to plant beautiful, yet dense and prickly, flowering plants outside our windows, doors, and entryways. The gate to my backyard is an obvious point of vulnerability. Predictably, two hearty and thorny bushes—a large bougainvillea and pyracantha bush—were planted on either side of it. I have permitted each bush to grow unchecked, so now the bushes intersect in places, forming a lovely and thorny archway above and around the gate. What was initially a four-and-a-half-foot-wide entryway is now half the size. It may not seem like a lot, but it is still one more obstacle to deter a would-be intruder. Using plants to beautify windows, gates, and entryways is certainly a familiar landscaping technique, but this aesthetic preference may stem from the fact that concealed entryways have, throughout pre-human and human history, been the safest. Our aesthetic appreciation of flora might be intimately tied to our ancient need for security.

A good security system will be multi-faceted: if one line of defense fails, there will be others to contend with. Let's assume the thorny bushes on either side of my gate are not enough to discourage an intruder and he opens the gate. At present the gate features a harsh squeak. I considered lubricating it until I realized I had a rather handy (and free) alarm system. Part of what creates sensitivity to certain sounds is our anticipation of them. I recall several occasions when my son Wilder was sleeping in the other room and either Kim or I thought we heard him crying, only to rush in and find him fast asleep.

I wonder to what extent the human auditory system has evolved to detect some sounds more readily than others. Just as we are prepared to detect survival-relevant sights, the same may be true for survival-relevant sounds. One question we could ask is if certain sounds have greater survival relevance than others—and if so, what those sounds might have in common. How is it that hearing certain sounds and not others helped (and helps) us survive? Perhaps the answer has something to do with the sonic mean, in which case it would have been—and still is—helpful to hear and make sounds with frequencies distinct from the mean. For example, elephants use a form of infrasonic communication that enables them to communicate over vast distances of savanna unbeknownst to humans and other animals. This, along with my sensitivity to the squeak of the gate, raises more questions: how do we normally interpret higher frequency sounds—and why do animals make them?

I recall the mockingbird in my orange tree, singing in full-throated ease at 2:00 a.m. Male mockingbirds that have not had success attracting a mate will sometimes sing late at night, when there are fewer sounds—including the songs of other males—with which to compete. Birds and other animals are known to sing as a way of demarcating territory. Wolves use their "songs" or howls for various reasons, but one is to alert other wolves to their presence, thereby establishing their territory and avoiding conflict. Whether animals make sounds to attract a mate or to signal territory, it helps if one can rise above the din (I think of the two little boys I saw the other day at the store, making a game of out-shouting the other). I think there is at least one other explanation for my sensitivity to the squeak of the gate and other high-pitched sounds, and that is because—as in the case of Wilder's phantom crying—we tend to associate high-pitched sounds with alarm, distress, and excitement.

Last week I was out back caring for our tomato plants when I heard what sounded like the grating noise of a rusty hinge coming from the shamel ash tree in my front yard. I had heard the sound before and recognized it as a mockingbird. As I listened I heard the same sound coming from Iliya's ficus tree across the street. Curious, I walked slowly toward the ash tree to see if I could get a look at the maker of this obnoxious noise. Deep in the middle of the tree sat a rather pathetic-looking fledgling mockingbird. The absence of long tail and primary feathers; the presence of downy, speckled chest-feathers; and a very large, yellow-rimmed mouth were diagnostic. A moment later, mom or dad flew into the tree with an insect-laden beak and proceeded to feed the flailing fledgling, which flapped its tiny wings and increased the rate and volume of its noise.

After finding this sonic equivalent to Chinese water torture, I thought about the different species of baby birds I have seen, including robins, hummingbirds, and even eagle chicks. Although varying in size and coloration, these baby birds have several things in common, including their downy appearance, high-pitched cries, and disproportionately large, colorful mouths. This interspecific uniformity is an example of an epigenetic rule where normally developing chicks, regardless of species, have large mouths to help ensure the conveyance of food from parent to offspring. As we saw in the case of security, the more we can do to ensure certain outcomes, the better. But having a large and colorful orifice into which food may easily be stuffed is only useful to the extent that the parent is encouraged to return with food. This may explain why mockingbirds have developed such an effective, attention-getting, sonic complement to their enormous mouths.

Mockingbirds are beautiful and elegant birds. Because of their delicacy, I am often surprised by their willingness to dive-bomb our cats, sometimes, it seems, for the sheer joy of seeing them duck in fear. This element of playfulness seems to disappear, however, when there are offspring involved. After three or four days of watching the mockingbirds carry out their parental responsibilities, I realized that there were two fledgling chicks, each in a separate tree. Except for a call or two, the chicks were silent until the parents returned with food or alighted nearby—then the noise-making would begin. This homeless strategy of rearing young piqued my curiosity. I have never seen a mockingbird nest, but given the bird's size, I suspect they make comparatively small nests. Even if the adults had built a nest twice their size, I can certainly see how difficult it would be to accommodate two little fatties such as these fledglings. Still, these birds seem a bit young to have left the nest. Perhaps the nest was disrupted or destroyed. Whatever the case, the mockingbird seems to have found effective ways of dealing with this situation.

This pair of fledglings is in a kind of developmental limbo: on the one wing, they're too big for the nest, and on the other wing, they're weak fliers. Nests are safe—they tend to be well-hidden and easily defended. With its offspring contained in a nest, an adult mockingbird can easily protect them. Now that they have left the nest and are more vulnerable, the reverse logic is used: by quietly sitting in separate trees, the chicks presumably stand a better chance of going undetected. In the event of an attack, the adult mockingbirds would only have one chick to defend, the other being safely hidden away in some other location. If attacked, a chick's inexperience and poor flying ability might handicap the adults' attempt to protect it. This apparent disadvantage would seem magnified by the presence of multiple chicks. Chick dispersal seems to be an effective defensive strategy, but it could also function as an insurance policy. Assuming one chick is lost to predation, the parents would still have the remaining chick. Thus, what might seem like parental negligence could be, given the challenges faced by mockingbirds, actually just common sense.

In fact, based on my observations of this drama as it often plays out between the mockingbirds and my youngest cat, Bella Jean, I would argue that mockingbirds are among the most loyal parents I've seen. They often put themselves in harm's way by flying dangerously close to the cat and alighting nearby. Once they've gotten Bella Jean's attention, they give her a good tongue-lashing. Both the feeding call and the alarm call of a mockingbird are hard on the ears, but the alarm call bursts into the eardrum, whereas the feeding call scrapes across it. The feeding call is comparatively discreet and difficult to trace, which must be useful in helping the chick remain undetected. In order for the alarm call to be effective, however, it must be loud, clear, and easily traced to an angry parent. Otherwise, the strategy of distracting the cat would fail. Bella Jean's hearing is very sensitive compared to mine, so I can only imagine how unpleasant the sound must be to her.

Our extreme response to the sounds of ambulance and police sirens highlights the significance of high-pitched sounds—as does the broad range of human vocal abilities and how they affect us. In terms of sounds that signify and elicit excitement, alarm, and distress, my son Wilder's vocal abilities are illustrative. Few sounds please parents more than the voices of their contented toddlers. Thus, whenever Wilder, our own little Emperor of Ice Cream, delivers one of his pre-lingual, experimental orations, Kim and I lovingly hang on his every coo and babble. We tell him that we love his sweet little voice and encourage him to talk more. Wilder's fussing noises are sweet sounding, too, but they are tempered with minor urgency. If Kim and I do not respond appropriately to this gentle cue, the fussing will intensify into bursts of outrage. Although amusing, these outbursts are characterized by their maximum urgency. "Ok," he seems to be saying, "*this* is serious." Wilder will tolerate a laugh or two, but if his needs are not met immediately, his bursts of outrage transform into an all-out tantrum.

While each of these high-pitched sounds serves a specific purpose, each sound is similar in its importance to the survival of the signaler. I realize that the sound of a squeaky gate may not seem relevant to the cries of a mockingbird, the wail of sirens, or a human infant's litany of coos. So let me trace the squeak of the gate to an ancient repository of survival-relevant sounds. These sounds would have alerted our early ancestors and other animals to environmental dangers such as storms and predators; to the whereabouts of prey; and to the needs of offspring. Although the forms or the causes of the sounds may have changed (compare the ancient sound of angry bees with the recent sound of electricity flowing through wires), we tend to more readily detect sounds above or below the sonic mean because doing so helped us to survive and reproduce. A squeaky gate is not ADT Home Security. But my goal is not to compare the efficacy of ancient and modern alarm systems. Rather, it's to suggest that we are biologically prepared to deal with threats to ourselves, families, and dwellings—whether we are aware of it or not.

The notion of preparation takes us back to the beginning of this discussion and my different sleeping positions. As I said, I hear much better from my left ear than from my right ear. I also noted that at night, and without exception, I cannot sleep unless my left ear is exposed. However, during the day I have no trouble falling asleep with my "bad" or right ear exposed. The question arises: why would hearing well be less important during the day than at night? Perhaps the answer has something to do with the fact that there is simply less to hear at night. After all, like us, other creatures are asleep during the peak hours of night, and if they aren't asleep, they are usually quiet, either because they're hunting or trying to avoid predation. In any case, maybe it's because we don't expect to hear sounds that the sounds we do hear seem amplified both in volume and importance.

I owe this awareness in part to the mockingbird in my orange tree, which for the past few nights has sat outside our window and sung into the early hours of the morning. Although I am at the point now where I can usually tune him out, the other night I confused one of his calls for an alarm. I could not fall back to sleep right away, so I just lay there (on my side) and enjoyed the sounds of the night. Then I noticed some interesting differences between this bird's "night style" of singing and the singing style employed by birds during the day. Now that it is spring, when Kim, Wilder, and I take our daily walk through the neighborhood, we can expect to see at least ten mockingbirds perched high atop some tree, streetlight, or rooftop, where they all sing in earnest. The other day we stood beneath a mockingbird as it sang atop a streetlight. Their songs are so loud and clear, it is a wonder that some mockingbirds sing at night. But when I close my eyes I begin to appreciate the challenges of singing during the day.

At this very moment, there is a mockingbird singing from the rooftop across the street. His song floats in through my office window, and then it is gone, drowned by the *ratatatatata* of a two-propeller plane overhead. A few seconds later, singing again: a whistle borrowed from the curve-billed thrasher. Then a leaf-blower and the muffler of some tricked-out automobile. Did I just hear singing? I can't be sure. The mournful sounds of doves resting in the shade of our grapefruit tree. A dog barking. Ah, there is the mockingbird again. Just a short trill. Now children screaming and yelling in the schoolyard across the street. A verdin hopping and chip chip chipping in the bougainvillea. Above it all, the maniacal clucking and cackling of male grackles.

Out of this cacophony of competing sounds, the mockingbird's "day style" of singing emerges, which involves singing more or less continuously. To some extent the song succeeds in rising above the din, but it is diluted. Even so, by singing non-stop, the bird capitalizes on any windows of silence that may open. The "night style" of singing is just the opposite. Apart from the horn of a passing train, or the far-off sound of a siren, very little can be heard in the early morning hours. The noisy humans are asleep, and so the mockingbird in our orange tree has the airwaves to himself. The night style emerges. The bird sings a set of phrases for three or four seconds, then pauses, perhaps listening for any response. Then he sings another set and waits a little longer. Unlike his daytime counterpart, rarely will he sing for more than a few seconds at a time. The daytime singer will string together every song he has, but the night singer is a minimalist. His songs cut deep into the suburban silence and into my dreams.

Our sonic environment depends on what and how we hear. When our eyes close for the night, our ears take over and keep us connected to the world. While background noise affects our detection of anomalous sounds, in a nocturnal setting the auditory system temporarily replaces the visual system as the primary sense. Just as the eyes function best in certain light levels, so might the ears have developed a heightened sensitivity at night, one that would offset the disadvantage of having compromised or useless eyesight. Throughout pre-human and human history the majority of threats and hazards occurred during the night.

v. Leave Your Flowers by the Door

As much as I abhor human violence, I would kill anyone who attempted to seriously harm my family. I haven't been in a fight since I was thirteen years old, but I am prepared to commit homicide. This is an odd thing for a gentleman to concede. And yet I cannot think of another circumstance where killing another human being would be considered more appropriate or, if not appropriate, at least understandable. If this seems extreme, consider the custom of respecting the homeowner while in his house, and the strict protocols involved when entering someone else's home. Serious conflict may arise if we fail to honor these deeply ingrained behaviors. For if a person cannot feel safe and secure in his dwelling, he cannot feel safe anywhere.

When someone disrespects or attempts to harm us in the sanctuary of our home, the ancient tapes start playing. It is as though one eye (or temporally proximate lens) sees the transgression of the moment (a dinner guest making a pass at your spouse), while the other lens projects Pleistocene scenes from deep human history, a time when things would surely not have ended well for the interloper. Indeed, they may still not. That transgressions of this kind are informed by an ancient intensity may help to explain why even the most seemingly minor infractions can sometimes lead to the most heinous acts of violence. Most of us have developed strategies for avoiding or diffusing conflict and for living peacefully (instead of assaulting the hypothetical dinner guest, perhaps I simply tell him he needs to leave). If confronted with violence in proximity to their refuge, however, animals will sometimes fight to the death. For inside the refuge lies one's most precious booty, belongings, mate, and offspring.

If my observations of other animals, as well as my own experiences, have taught me anything it is that prevention is the key to protection. Since his birth, Wilder has slept with Kim and me in our bed. Kim and I enjoy having him sleep with us because we can watch over, feed, and soothe him should he wake disquieted. This seemed reasonable to us—until we met with Wilder's pediatrician: she thinks he ought to be sleeping in his own bed "yesterday." Kim and I agree that Wilder would be better off sleeping in his own bed, if not according to the doctor's timeline, then within the next few months, when he can eat other foods besides his mother's breast milk and can therefore sleep through the night without becoming hungry. In order to understand why we might resist moving him into his own room, I've pondered how the Desert Archaic peoples (DAP) might have addressed the issue.

Based on what I know about DAP's social behavior, it seems unlikely that parents and elder tribal members would have designated separate sleeping quarters for their children—certainly not during the first years of life, when human offspring are more vulnerable. Surely DAP and other early human parents felt a corresponding need to sleep alongside their children. Maybe they didn't feel this need in the same way that we do now, but their instinct must have told them that—in addition to encouraging the all-important bond between parent(s) and offspring—it made good survival sense to share a bed until the child could contribute to his own safety by hiding, fleeing, climbing, or sounding the alarm if threatened or in need. Wilder's pediatrician would no doubt be skeptical of this view. In fairness to her, research does show that parent/child bed sharing can sometimes be dangerous, and most children no longer face the threat of being carried off by hyenas. But there are still predators—those that are real and those that haunt the Pleistocene mind. Surely current medicine isn't so advanced that it has transcended the relevance of deep history and the history of the human family.

For now, Wilder will sleep with us, but when we do move him to his own room, we will have to ensure his safety in other ways. Despite the fact that we live in a relatively safe neighborhood (also an important part of the habitat selection process, which I will discuss shortly), Kim began pondering this issue shortly after Wilder was conceived. Of particular concern was where, given the floor plan of our home, Wilder's room should be. Our home has three bedrooms: the master bedroom is on the west side of the house, and two adjacent rooms are on the east side. The southernmost room faces the street and is (by default) Wilder's room. (Let the record show I chose the northernmost room as a study a year before Wilder was conceived!) By the time Wilder is ready to sleep in his own room, I will be evicted from my digs because Kim has decided that my office would be safer.

I agree, and I also like the idea that Wilder will enjoy the view of the eastern portion of our backyard, a lovely little oasis of trees, plants, and flowering bushes. More importantly—and this is Kim's point—the area is separated from the front yard (and therefore from the street) by a six-foot, cinder-block wall that has large (and thorny) bougainvillea bushes growing along its outer, southern side. On the other hand, Wilder's current room faces the street, which could be a problem. When the time comes, I will defer to Kim's discretion. Interestingly, Kim and I are not alone in our concern for security: on our daily walk we pass one hundred houses, each with features that help to offset its exposure and vulnerability. And my guess is that the occupants of the homes went through a similar, albeit perhaps unconscious, process when deciding on rooms for their children.

For Jay Appleton, father of habitat theory, the best refuge features proximity to resources or prospects while allowing inhabitants to see without being seen. Like other Native American tribes whose cliff dwellings are found throughout the Southwest, early desert peoples accomplished this in their mountain retreats. However, we modern-day suburbanites have had to find other ways to shield ourselves and to create privacy while living in the midst of thousands of other people. Fortunately for Kim, Wilder and me, although the fellow who originally owned our home may not have known about habitat theory, he (or perhaps the landscape architect) still took certain measures that augmented our sense of security, and that would, in effect, enable us to see without being seen.

For one thing, the past owner had the yard professionally landscaped about seven years ago, so by now all the plants, bushes, and trees—including a twenty-five foot tall jacaranda tree outside Wilder's bedroom window—are well established, and create a perimeter around the house. Apart from being elegant and featuring large purple flowers when in bloom, the jacaranda has the benefit of shading Wilder's room from the sun and blocking the curious (or malicious) stares of passers-by. In the event the tree does not quell a potential intruder's curiosity, two rather bushy bottlebrush bushes just outside Wilder's bedroom window provide the next line of defense. Like the jacaranda, these bushes provide both a visual and physical barrier to the outside world.

Similarly, each window on the east, south, and west side of the house has a sunscreen and some form of botanical barrier. During the day, these screens enable us to see outside while making it impossible to see inside the room. If, however, the blinds are not closed while the bedroom light is on at night, the opposite effect occurs: we can't see out but people outside can see in. Still, the sunscreen and bushes are useful deterrents should an intruder decide to come in for a closer look, not to mention that an intruder would also have to contend with the window itself. For whatever reason(s)—instinct, aesthetics, budget limitations, error—the contractor who built our home fashioned the room with only one window. That he didn't also construct a window in the east-facing wall to give a view of that part of the front yard, which is quite lovely, illustrates how safety and security usually trump aesthetics. Sometimes the two work in tandem. The goal is to find a balance, and that is what the builder did by including the one window.

Wilder's current room is, therefore, relatively well fortified, but not well enough to change Kim's mind. Call it paternal instinct or common sense or a desire for marital harmony, but I tend to agree with her. When compared to the fortification of the study—the cinder-block wall, the thorny bougainvilleas, the massive potato bush beneath the window and the closer proximity to our bedroom—the choice is clear. The black widow and the fledgling mockingbirds come to mind: better to have the nursery deeper in the dwelling.

Just the other day while planting a gardenia, I sank a shovel into one of the many ant colonies around our house. I must have penetrated the nursery because worker ants worked frantically to secure their cache of rice-sized eggs stored deep in the ground. Humans are not alone in their need to select the safest place possible for their offspring.

vi. Caves: Affordable Housing

In the past, refuges that were safer and more secure than others would have been selected over less-fortified dwellings. This doesn't mean that other places were ignored; it just means they were used in ways that were not contingent on a high degree of safety. The Gold Canyon site is a case in point. Although the knoll or "island" does have the advantage of offering far-reaching views of the area—very important for the detection of enemies, inclement weather, predators, and prey—it would not be easily protected from predators or from enemies in the event of a surprise attack. Given this vulnerability, the Desert Archaic peoples likely used this site on a short-term basis to regroup and to attend to daily needs such as the preparation of foods that had been hunted or gathered from the surrounding desert. The many grinding holes found atop the knoll and along its north (and therefore shaded) side suggest that food was prepared here throughout the year. I have no way of knowing whether they spent nights here or not, but if we have anything in common (and I believe that we do), they would not have made a habit out of sleeping here.

The site features a small cave on its lower northeast side, but instead of being elevated above the ground, which would make it less prone to invasion, the cave is on ground level. (If Kim and I were to remove the south wall of our front room, I suspect we would have created a similar effect.) I have sat in this little cave as a reprieve from the sun, and I think the DAP most certainly did, too. Maybe they took naps here while others in their party sat atop the knoll and kept a lookout. But with the nearing of night I suspect they headed for higher country, perhaps to a place more like the one Appleton had in mind.

I have visited five different sites within a five-mile radius, including three different caves. The caves vary in size from "extended" to "multi-family" dwellings. Cave A faces west and could shelter a dozen individuals. Cave B, which faces east and is located in the canyon just west of Cave A, is massive and could easily accommodate one hundred individuals. Cave C is also massive, but instead of being one giant cave, it is a series of smaller caves protected by an enormous overhang of volcanic rock. Cave C is located two canyons west of Cave B and has a southern exposure. How many people each cave accommodated varied, but what do not vary are the characteristics that make these caves appealing refuges.

Recalling Appleton's criteria, we would expect to find refuges in proximity to resources or prospects (another word for "opportunity"), such as water, edible plants, and animals. Not surprisingly, then, each cave is indeed positioned within a hundred yards or less of a riparian area or waterway. Although these waterways may go many months without water, they still feature relatively abundant vegetation, which attracts important prey species such as rabbits, quail, mule deer and javelina. These waterways also provide access to higher elevations, which is where bighorn sheep—a revered prey species—are found.

Each cave is located high above the desert floor. Cave A is situated at the lowest elevation, but like the other two caves, which are considerably farther up the mountain, it features an unimpeded view of the surrounding area. The range of view increases when, like peering from a tower, one stands atop the rock formations that comprise each of the caves. Having stood inside and atop the caves myself, I can appreciate how difficult it would have been (and still is) to approach without being seen. This brings to mind the familiar saying, "There is safety in numbers." Assuming the Sonoran Desert was significantly more dangerous in the past, cohabitating with other people clearly had certain advantages, especially when it came to protection, obtaining food, and detecting danger.

The appeal caves have for us is not simple. They were and are potential sources of danger, particularly with respect to large prey animals or other humans that might be found within them. Curiosity and the equally powerful drive for food and resources that the cave's occupant may provide increased our fascination. Each year, thousands of people pay to visit major cave systems throughout the country, and spelunking continues to be a popular form of recreation for the more adventurous among us.

The sheer effort required to reach the sites in my study area is suggestive of a profound attraction to caves. The popular idea that we climb a mountain "because it is there" is true to a degree, but we do not risk life and limb for nothing. From the perspective of the Pleistocene mind, there is nothing arbitrary or whimsical about the decision to climb a mountain or, in this case, hike to a cave.

As anyone knows who has ever stood atop a mountain, the view is panoramic. For a species whose life depends on assessing and exploiting resources, locating danger, and finding places of refuge, seeing for miles and miles in every direction is helpful indeed. But our willingness to ascend mountaintops and explore dangerous caves isn't the only indication of this deep-seated relationship with the environment. Our capacity for route finding and selection, and our psychological responses to caves and other significant landscape features are significant as well.

Of all the caves in my study area, Cave C—a series of parallel caves united by an enormous overhang of volcanic rock—is by far the most attractive and the most difficult to reach. Cathedral-like, the cave can be seen from miles away. Given its conspicuous nature and its apparent inaccessibility, I had long wondered if the cave would yield signs of human habitation. Inaccessibility is a desirable characteristic for a refuge, but after a certain point it would become too costly for an inhabitant (at least a human inhabitant) to reside there. Assuming daily caloric intake was considerably less, and at times unreliable, for the Desert Archaic peoples and their successors, why they would choose a dwelling that apparently cost them so much energy to reach just didn't make sense.

However, in other ways the cave seemed ideal. It is adjacent to a large wash, and is strategically located between the desert floor and the upper reaches of the Superstition Mountains. Each elevation may have offered different resources: the low desert is richer in edible plants and small game animals such as rabbits and quail. In addition to their far-reaching views, the higher elevations are still home to bighorn sheep, an animal depicted with great care on rocks throughout the desert. Given these apparently conflicting variables, I had another puzzle on my hands. How could a place seem so right and so wrong at the same time? Something had to tip the aversion/attraction scale one way or the other. I worked through the predictions of habitat theory and realized that I had concluded the cave was largely inaccessible before identifying the route(s) used to reach it.

Reaching the cave was a visually daunting task—I was about to find out if it were physically difficult as well. I took off my backpack, put down my hiking staff, pulled up a rock, and sat in the shade of a palo verde tree to do a little route finding. Within a few moments I identified three possible routes to the cave, the directness of which decreased with each route. The first route involved hiking up a wash and then climbing an approximately 150-foot cliff/waterfall. The second route was east of the wash and would require scaling thirty feet up a cliff face to a band of vegetation that, if followed, would lead to the cave. The cliff face bears traces of foot and hand holds. The third route was west of the wash and follows a ridgeline to a steep mountain slope, and then traverses east beneath a massive cliff wall toward the cave. "Tell me:" writes Theodore Roethke, "Which is the way I take; / Out of what door do I go, / Where and to whom?" Roethke scholars will forgive me if, in my effort to explore the adaptive significance of way finding, I add one additional question to the poem: Why is the way the way I take?

vii. The Way I Take

The ability to identify potential resources would have been paramount to our individual genetic survival. Our ability to capitalize on available prospects would have involved knowing how to navigate the landscape. It is one thing to identify a vantage point high up on the mountain. But it is quite another to decide how best to get there. The majority of modern humans may not make the effort to reach the mountaintop because the necessity of doing so has long passed—but the attraction remains. Humans and other animals have been navigating their environments for millennia. Given this long-term relationship between our genes and the environment, it is not unreasonable to suppose that over the years we have developed an innate sensitivity to environmental cues affecting our ability to reach a prospect.

A parallel is the daily commute many of us make to our places of employment, which could be considered the modern form of prospect. When Kim and I first moved to Gilbert, I went through a process of route finding until I finally found the streets that got me to work the quickest and the safest. I suspect my early human counterparts used similar criteria when they began navigating their way to the mountain caves. Doing otherwise would be a waste of precious energy and would expose travelers to hazards. Obviously, early humans did not have to deal with rush hour traffic, but throughout human history humans have, when traveling, kept a watchful eye on the rising and sinking sun.

Compared to the onset of night, the rising sun inspires a sense of relief for many creatures. For diurnal creatures such as us, night is the hard time. We have even developed the folk idea that everything "bad"—including various illnesses, both mental and physical—worsens at night. Thus, our need to find shelter, or to return home by dark, is profound. Navigating difficult terrain in the dark is risky, but night is also the time when predators become most active. Most humans are no longer threatened by roaming hyenas, lions, and other dangerous animals. The majority of natural environments, particularly those tangential to human civilization, have largely been cleared of threats to human life. Even still, for many of us the need we feel to return to our refuge is rivaled only by the anxiety we encounter, however briefly, when the sun goes down in the wild.

Gordon Orians and J. H. Heerwagen have examined hundreds of landscape paintings in an effort to explore the adaptive explanation for our responses to light and shadow. They note, "Throughout human history, night was a dangerous time ...The setting sun would have been a strong signal to return to a safe place for the night." Their survey of landscape paintings bears this prediction: "Not surprisingly, artists incorporate more negative responses to sunsets than to sunrises in their paintings. Compared with paintings of sunrises, paintings of sunsets have, on average, fewer people, and they are closer to a refuge than are people in paintings of sunrises. In addition, people that are far from a refuge in sunset paintings may be portrayed as anxious."

When my friend Greg and I hiked to Cave C, we were keenly aware of finding a route that was safe but that would not leave us stranded on the mountain at dark. We left the Gold Canyon Trail and headed for the ridgeline, where we discovered a well-traveled animal path, a route that also may have been used by people long ago. Is it possible that different species use the same criteria when selecting a route? Animal paths tend to be spatially conservative compared to the paths we associate with modern humans. But early human trails probably did not make a discernibly greater impact on the land than did other animal trails, in which case it is quite likely that early peoples also traveled these thin paths with some regularity. As Greg and I hiked along the ridgeline, I wondered why animals prefer exposed and often indirect ridgelines when traveling.

Greg knew I had been thinking a lot about habitat theory so he did not object when, on our return from the cave, I stopped on the ridge and asked him some questions. After establishing that he shared my sense of urgency at the sight of long shadows, we noticed factors that might affect our preference for the ridgeline. Perhaps the most obvious reason for preferring ridgelines to other routes (e.g., riparian areas or corridors) is because of their views. If Greg and I were walking just ten feet below the ridgeline, our view of the surrounding area would be reduced by half. In addition to preventing us from identifying prospects in the area, this reduction in visibility might enable a predator, hostile human, or even a storm to approach without being detected. Walking the ridgeline would therefore be much safer than using a low-lying route.

Our purpose for moving through a landscape affects our

navigational decisions. For example, both traveling from one place to another (e.g., the low desert to a mountain refuge) and foraging require us to move through a landscape. However much these reasons for travel might overlap, they each pose specific challenges and require us to respond appropriately to different aspects of the environment. Standing on the ridgeline, Greg and I could see a wash that began at the Gold Canyon Trail and ran, much like a natural staircase, all the way up the mountain. Heavy vegetation grew in places along the wash, but surely it wasn't so thick that we couldn't have worked around it. Seems like a reasonable compromise, which makes me wonder why we hadn't gone that way instead of making the extra effort to ascend the ridgeline.

Riparian areas typically offer a broader array of vegetation than outlying areas, attracting other animals and birds that, together with the plants themselves, would have comprised the menu. By the same logic, we too would have been on the menu. Therefore, we would have been especially vigilant when moving through the often-dense vegetation in search of food and this would have affected our movements. I think of browsing deer, nibbling and listening, nibbling and then moving a few feet and listening again. Desert Archaic peoples employed a similar technique when foraging for food, but when it came time for traveling to the next feeding area, camp, or shelter, personal experience tells me they made a beeline instead of the fits-and-starts of browsing deer.

As we all know, traveling is dangerous. Just last March, Kim, Wilder, and I were driving across the Mogollon Rim on a return trip from New Mexico. That part of the state is known as "elk country," so every few miles there was a sign or a series of signs warning drivers to be alert. Colliding with an elk is a bizarre notion. It's only been within the last two hundred years or so—with the advent and the eventual refinement of the combustion engine—that we've even been able to *imagine* colliding with an elk. Before that time we'd be lucky to *catch* an elk, let alone crash into one with enough force to kill the elk and ourselves. And how old is the human species? It depends on whom you ask. Personally, I hold with those who place the emergence of modern *Homo sapiens* between 150,000 to 200,000 BP. In any case, we're talking about a long time to become set in our ways, a notion that is supported by the fact that drivers are inundated with warning signs.

Colliding with elk or other animals is a very recent phenomenon, but travel of any kind has always exposed us to hazards. We are vulnerable away from our home turf. Consequently, we may feel a sense of great urgency to reach our destinations, places which, however distant, are familiar and secure. Even now, when Kim and I travel four hours by car to the White Mountains, I often have to check my Pleistocene mind and remind myself that it's alright to stop along the way, a tendency I seem to share with many of my male friends and colleagues. Imagine the urgency we would have felt when traveling on foot! Imagine the peril! To avoid becoming a target, the incentive to keep moving would have been strong. Thus, if an animal's purpose were to get from point A to point B, it would make very little sense to chance a hostile encounter by traveling through an area that is typically used for feeding purposes, or by dilly-dallying once travel had begun. In an ancestral environment, safety would have been a constant concern for us. Certainly we would have benefited from modifying our defensive responses to differing situations (foraging vs. traveling) and topographical features (riparian areas vs. ridgelines).

Whenever I stand at the foot of the trailhead, or look at photographs I have taken of the area, and ponder the different ways of reaching Cave C, I am struck by the desert's deceptive appearance. Entire hills and rugged contours are blurred when viewed from afar. Washes dissolve against a blanket of green, which is really an assortment of cholla, prickly pear, succulents and other desert flora of which one must be constantly mindful. Riparian areas also appear easy to navigate, but when

I enter a wash (or any part of the desert, for that matter), I begin to appreciate the demands of the terrain, including boulders and tangles of thorny trees and bushes. Walking a wash requires careful consideration, where the search for a passage must be balanced with need for safety.

If finding food were my goal, I would accept the demands (and dangers) of wash walking as an unavoidable part of the process. But if my goal were to travel, I would leave the wash and head for the ridgeline. Like other wind-swept places, desert ridgelines are typically bare, with no significant obstacles. Also, the breezes make traveling in the hot sun a little more bearable. Just yesterday, I hiked along a ridge in the Superstitions—it had to be ten degrees cooler there than down in the lowlands. This difference in temperature is very important when traveling. Despite the fact that water is now so accessible, I am still wary of dehydration. I have a profound respect for the early inhabitants of the Sonoran Desert, for whom the acquisition of water must have been challenging. This is to say nothing of their footwear—another reason to choose the path of least resistance! All desert life has had to become biologically conservative. Javelina, coyote, mountain lion, human, whatever—it is important to reduce exertion and maintain body temperature: a nice breeze helps us do just that.

About half way to the cave, nearly a thousand feet up the mountain, there is a surprisingly level shelf that is protected from the wind and sun by an outcrop of rocks and a palo verde tree. I remember seeing the area from below and thinking it would be a good place to rest and enjoy the view. But I had to get there first. Although the ridgeline trail had dissolved once we reached the mountain slope, I encountered various animal signs as I made my way to the shelf. Again, the desert residents and I seemed to be selecting the same, albeit less distinct, route up the mountainside. This mountainside is already a dangerous place, but due to heavy rain, which had saturated the soil in the days leading up to the hike, it had become much more so. The shelf offered Greg and me a safe place to rest before we made our final push to the cave.

Among the vast collection of coyote, deer, and bighorn sheep scats, I found pottery shards scattered all about. It is wonderful to think that such different animals, each with its own niche in what Charles Darwin described as "the community of common descent," all found our way to this exact spot on our way through life. Although I did not attempt to take the other two routes I had identified, I suspect they would have cut at least an hour off the time it took Greg and me to reach the cave. More importantly, these alternative routes are keys to resolving the conflict between the cave's desirability and inaccessibility. Why might the early inhabitants have used multiple routes to the same place?

Orians has observed that our responses to "environmental cues vary with a person's age, social status, and physiological state." I've seen this variance in action whenever I've gone hiking with my mother or brother-in-law, whom I mention because they offer the most dramatic comparison. I've already recounted my mother's apparently irrational (Pleistocene?) fear of being lost even though we could see houses in the near distance. However, I have sometimes wondered if my brother-in-law isn't too relaxed, to the point of carelessness, when out in the desert. My mother was sixty-two, James was twenty-seven, and I was thirty-seven with a toddler son. Compared to James, I must seem like a bit of a worrywart. My mother would no doubt accuse me of not worrying enough. Whereas Greg, who is two years older than me with three small daughters, is similar to me in terms of the risks he's willing to take.

Of course not everyone in these age brackets behaves in accordance with the predictions of habitat theory. But in general, I think we would find consistency between peoples' stations in life and their responses to the environment. Perhaps at some later point I could test this hypothesis by providing different aged and sexed people with a photograph of Cave

C and the surrounding area. I could ask them to determine and chart routes to the cave in order of "best" to "worst" and then explain their criteria. I could also ask them to select the route they would take.

Habitat theory predicts that older people and women would, on average, find more route possibilities, but would still select the least strenuous and/or safest route. Older people would be less fit physiologically, whereas women would tend to select the route most suited to conveying or guiding children, whether they actually had children or not. On the other hand, I would predict that young males, who are notorious risk-takers, would identify fewer routes and might even select the route that most exposes them to risk. After all, it is mostly young men who join the armed forces, and who most often die in automobile and motorcycle accidents.

Thus my status as a thirty-seven-year-old male helps to explain my route selection. Although I am not quite in my physical prime, I could have as easily ascended the waterfall route as any twenty-year-old. And yet I did not choose that route or the other route like it, even though doing so would have been much less strenuous and time-consuming. Instead, I chose the route that exposed me to the least amount of risk. I'm sure there are multiple reasons for why this is, but remember that "social status" is one of the factors affecting our responses to environmental cues. Social status is complex. I am many things, including a new, first-time father. Perhaps in some societies natural selection would have favored males who, despite their powerful sex drives, avoided high-stakes risks once they had reproduced.

This is not to say, however, that the route Greg and I took was without risks. I know for a fact that Greg and I both felt anxious as we made our ascent. In order to understand this, it helped to think about the feeling itself. "Anxiousness" was a good start, but the description needed to be refined. After some reflection, I decided I was feeling "vulnerable" or "exposed" as I made my way up the mountainside. Compared to the feeling of anger, which is generally directed at something specific, the feeling of exposure is generalized, and prepares us for the greatest range of hazards, some of which have persisted since Pleistocene times, some of which have not. From the perspective of the Pleistocene mind, then, it does not matter that most large predators and hostile conspecifics are now extinct from the Superstition Mountains. What matters is that we were and are vulnerable to any number of threats—storms, falling rocks, predators, falling—when in the open. Greg and I were not only in the open, but we were also on a precipitous mountainside. Thanks to my feelings of vulnerability and exposure, I became more aware of my surroundings and focused on reaching a place of safety.

For the evolutionary psychologist, emotions and feelings are adaptive if they promote behaviors and responses that contribute to survival and, ultimately, reproductive success. What an interesting way of thinking about the role of "negative" emotions in our lives! Although we normally discourage so-called negative emotions in ourselves and in others, when we examine our responses through an adaptive lens, we find that our "negative" emotions are every bit as valuable as our "positive" emotions. In fact, when it comes right down to it, all normal emotions might be described as "adaptive" to the extent they promote survival.

But if I describe as "good" emotions that are considered "bad," or undesirable, no one will know what I am talking about. Imagine describing jealousy or rage as a "good" feeling. This will not do. We are social creatures, after all. We describe our feelings and emotions in an effort to understand them, to *know* them. But it would help if our descriptions were informed by an understanding of deep history. Most of us are content to describe the feeling itself, whereas the adaptive perspective is more concerned with understanding why the capacity for this kind of feeling emerged in the first place. One of the problems with evaluative or teleological thinking is that feelings are often reduced to "good" or "bad." Not a lot of understanding going on here, I'm afraid.

This may be acceptable from a social point of view, but this mindset may be self-defeating insofar as it makes for some unhappy, not very well-oriented human beings. Humans who are at odds with human nature are at odds with nature. They lash out against, or are indifferent to, a world they see as separate from them; a world incapable of enriching their lives. Humans have long been preoccupied with transcending the Earth and corporeal constraints. This ancient need is understandable: the world can seem alien when we do not know how we fit into it. Consequently, over the course of human history we have found ways to distinguish ourselves from a planet we did not understand in the first place. Humanity's struggle isn't so much with the Earth as it is with our own limited perspective of ourselves.

viii. A Case of Nerves

As someone who subscribes to the idea that humans did some serious evolving during Pleistocene times, I find it useful and exciting to speculate about how my psychological responses might be traced to our ancestral environment. This has not always been the case. Before I became reacquainted with my own Pleistocene mind, I would simply suffer, endure, or take my feelings for granted. What was important was that I had feelings. It didn't matter that there might actually be a reason for them beyond the obvious. All feelings are interesting to contemplate from an adaptive perspective, but I wonder about those feelings we normally describe as negative or painful or unpleasant. What is their survival relevance, and how might understanding that relevance help us become more in touch with ourselves as human beings? Although I can point to examples from every area of my life, some of the most fascinating responses have occurred in relationship to Kim, who is my wife, mate and, as of October 27, 2005, the mother of my child.

Kim and I do a fair amount of hiking and fishing together in the deserts and mountains of Arizona. Kim is a wonderful companion and fellow adventurer. But it has only been recently that I have learned to fully enjoy our outings. Before, our adventures were often very stressful. In fact, at times my feelings of distress became so pronounced I thought I might have some kind of problem. I was troubled by my apparent inability to relax and enjoy myself. Then I began to notice things. For instance, I did not feel stressed when in the wilderness with my friend Bert. As I thought about it more, I realized that I have *always* been more comfortable around males when in the wilderness. Does that mean I like Bert more than Kim? Or that I have more fun with Bert and my other male friends? Do I have issues? What's my problem? But I wasn't getting anywhere by asking these questions. If I could not avoid them, I suffered these feelings like rocks in my boots. At the time and because of my perspective, I had no other choice.

I think we all suffer (and inflict suffering on others) because of our perspectives. But that is the nature of teleological thought: any behavior that is not ideal is considered a deficiency or a "personal problem," which is then codified as aberrant or as having no useful connection to the larger phenomenon of human nature and experience. In a climate that, ironically, has relatively little tolerance for and understanding of so-called "abnormal" behavior, we become our own private freak shows. It was only later, after I had begun phasing out evaluative thinking and replacing it with an adaptationist approach, that I began to understand the significance of my apparently "deficient" response.

This recognition began when I inquired into the survival significance of the response and why I might have evolved to feel one way in the presence of males and another in the presence of females when in the wilderness. Although it's possible to approach this question by working from the top down, I sometimes find it easier to understand the dynamics of what is by speculating about the dynamics of what was. So, working from the bottom up, what would it have meant to hike with Bert (or a capable male) as opposed to Kim (or a capable female) at some point during early human history? And how might those differences explain my current responses? Perhaps the most obvious difference is that Kim is my mate and the mother of my child. It could be that the anxiousness I feel when in the wilderness with Kim is my body's way of preparing to defend her—the potential and actual mother of my offspring—in the event of danger. If so, then males with this protective tendency would be more likely to have and to keep offspring. This might also help us to understand why males' protective instinct persists, and is still highly valued today.

This explains why I felt more at ease with Bert. I am not sure how great a role my relationship with Bert played in my response, or to what degree his status as a non-relative affected the comparative lack of responsibility I felt toward him. Perhaps I would have felt differently if he were my brother or father instead of my friend. It's nothing personal, but from a genetic point of view it would have been more important to protect family than non-family. Our relationships with people affect how committed we are to sticking out our necks for them. But I don't think it's just that. Another reason why I don't think I felt as concerned with protecting Bert is because as a capable male, Bert can fend for himself. Thus there is an enormous difference between the responsibility we feel to protect ourselves and to protect someone else, especially if that someone else is our mate and the mother of our offspring.

Moreover, a male and female are generally much easier to overpower than are two males. I know there are women in the world who are capable of defending themselves, but if Kim and I were attacked I would feel compelled to protect her while at the same time having to protect myself. I get anxious just thinking about it! Conversely, two adult males constitute a formidable force, like two male lions patrolling their territory in the Serengeti. Oh that I were that regal and awesome! As it turns out, a little anxiousness helps us defend our interests—what we typically perceive as a deficiency or personal problem may actually be an advantage in disguise. For although much of today's environment differs from the challenges faced in our ancestral environment, it still makes survival sense to be alert and to guard the well-being of our mates and companions when in the wilderness.

When I was on the mountainside with Greg, I did not feel the need to guard his safety and could focus on myself. It's true that I was a bit anxious, but this feeling was inspired by being on a steep mountainside, where I was exposed to actual perils. I know my feelings would intensify if I were joined by Kim and Wilder. If there is an agoraphobic continuum that ranges from debilitating to healthy, I would say I have a healthy respect for open places. Still, I was relieved when Greg and I had ascended the mountainside and began our traverse east toward the cave. It had been hard work reaching this point. Greg was a few minutes behind me so I took the opportunity to rest. I stood beneath an approximately thousand-foot cliff and scanned the area. The path there was well traveled. The loose dirt was pocked with hoof prints and clusters of dark droppings. Clearly this was a popular route, so I was only mildly surprised when, after I heard rocks falling, I peered across the cliff and saw a ewe looking at me. She stood about forty yards away, still as stone except for the occasional blink of her eyes. I whispered loudly to Greg, but by the time he reached me, she had disappeared around the corner. I guessed she had not seen our kind in the area for some time.

Bighorn sheep sightings are rare in the Superstitions, mainly because the sheep spend the majority of their lives high in the mountains. But over the last three years I've seen ten bighorn—rams and ewes—at low and high elevations. Even as a native Arizonan, Greg hadn't seen one, so he was excited by the possibility once we rounded the cliff corner. We got there just in time to see two bighorns cross the wash and make their way to the cliffs on the other side. Once there, they stopped and looked back at us. Greg and I took turns glassing them. What marvelous creatures.

But the sheep were not the only remarkable spectacle: Greg and I now stood at the opening of an enormous upside-down bowl. A gigantic chunk of rock had broken away from the mountain, leaving a crescent-shaped wall several hundred feet high. And there, at the base of the bowl, were the caves, still partly enshrouded by afternoon shadow. I don't think more than three or four minutes had passed, but when I looked back at the cliffs the bighorn were gone. I'd say they disappeared into thin air, but that would do no justice to how these exquisite animals navigate this rugged and dangerous environment.

Landscapes are complex places, and they can evoke an array of differing psychological responses from one moment to the next. At least that was my experience as Greg and I hiked our way up the mountain and deeper into the bowl. We had made good time getting to this point, a fact that gave some credence to my theory that the route we chose was used by the more conservative members of early human society. The sun was just west of one o'clock. We had several hours of daylight left, but oddly I still felt like we were running out of time. Although I did not realize it then, I think I felt conflicted by the bowl's different dimensions of light and dark. For although it was about 1:30 p.m., the cliffs cast long deep shadows that confounded my eyesight and triggered my sense of night-dread.

I felt caught in a kind of temporal limbo. But the caves beckoned; and so on we went, higher up the mountain and deeper into the bowl. We soon came to a ledge of rock that had to be climbed if we were to continue. The ledge was very simple to ascend, but by this point I had let my Pleistocene mind get the better of me. In fact, if Greg had not overcome his own case of nerves (which he later reported once we had reached the car) and scrambled up the ledge, I may have turned back right then and there, even though the caves were no more than a hundred yards away. Caves be damned!

Based on my experience of many landscapes, our responses to places are manifold. That is, we may have either a negative, neutral, or positive reaction to an environment, but each reaction is comprised of several variables. I can think of several aversive qualities, not because there weren't also attractive qualities, but because the aversive elements have greater survival relevance and demand immediate attention. I can afford to momentarily ignore a clear pool of water from which I plan to drink. I cannot afford to ignore the areas around the pool where predators may be hiding. So even though I was excited to have finally made it to the site, I was feeling a bit guarded for at least a few reasons, beginning with my unfamiliarity with it. Another factor was the latent and immediate awareness that where there are bighorn, there is possibly their major predator, the mountain lion. The contrast of light and shadow was yet another. But I still forged ahead, rationally aware of the time and that I need not worry about getting caught on the mountain in the dark. I was not of two minds, however, when it came to the feelings inspired by the spatial dynamics of the bowl. My minds talked it over and decided that I was feeling a twinge of agoraphobia's sister emotion, claustrophobia.

Claustrophobia is often such an extreme response that it is difficult for us to appreciate the usefulness of its milder manifestations. We may not be claustrophobic, the inability to escape from either an open or closed space strikes a chord in us. If the goal is to understand our responses—however severe or mild—it helps to think about why we have (and had) these fears in the first place. Let's assume I am anxious in enclosed spaces and am later diagnosed with mild claustrophobia. All this really tells me is that I am somewhat fearful of closed spaces. Presumably, the next step would be to understand why. Perhaps with the help of Dr. Phil I decide that the reason I am fearful of closed spaces is because when I was an infant and would cry my mother would put me in the closet with the shoes. Thanks, Dr. Phil! Nice shoes! Case closed. Well, not quite.

One of the goals of any psychological diagnosis is attribution. In this scenario, the "patient" gains control over his fear by identifying its apparent origin. But what if this attribution is only part of the story? What if the closet incident were an effect rather than a cause? What if I could go even further back in my past and link this fear (and my love of footwear) to humans who lived thousands of years ago and who in some sense made my life possible because they had a fear of, and therefore avoided, being trapped? Wouldn't I then be in a better position to address and use my fear and be empowered by it? Phobias *can* be debilitating. But the majority of modern humans is probably very similar to the majority of early humans and other animals, for whom a healthy dose of fear of enclosed spaces would have had, and still does have a direct impact on survival.

Part of what makes it difficult for us to appreciate our connections to the past is that our responses often appear incongruous with what is happening in the moment. It does not occur to us that our responses, while inspired by the moment, are rooted in the deep black soil of the human species. Lives are momentary but responses are ancient. This is why our hearts pound as much at the perception of danger as they do to actual danger. The fact that our responses often bear no obvious relationship to the experience of the moment tells me that our understanding of perception (and of the moment, for that matter) may be due for some revision.

We could begin by making the distinction that sociobiologists and others working in the evolutionary sciences make between proximate or immediate causes and ultimate or historical causes for behavior. My time in the closet provides an example of a proximate cause for my fear of enclosed spaces, but the reason I have the fear in the first place is because early humans had it and passed the genetic foundations of the response to me over hundreds of thousands of generations. The fear remains part of human nature to this day because historically it helped us stay out of trouble.

I don't mean to suggest it is possible to separate the proximate and the ultimate except with language, but merely to point out that millions of years are brought to bear on our experience of any given moment. Otherwise we could not possibly hope to explain the presence and the intensity of these feelings, especially when they often do not have obvious inspiration in the present environment. I worry that Wilder and other children aren't going to have the same opportunities to know themselves by learning about the environment. But I am heartened by the knowledge that there are a lot of people who are working to reverse our destructive behaviors. We can also learn a great deal from archeologists, physical anthropologists, and detectives of all kinds who can help us understand the importance of the integrity of place and of the need to preserve it. How else will we know how things have been with us? If we do not leave the environment intact, we will have lost the wherewithal to ask questions about and learn our true place in the world. We will have lost our inspiration.

ix. Song of the Cave Trance

Although we may tend to generalize when describing our responses to different environments, a little analysis reveals the richness of our feelings toward landscapes. Given how I was feeling when I stood below those caves with the hankering to turn tail, I imagine someone could ask why I even bother going to these places if they inspire such unpleasant feelings. It's a fair question, but remember that our responses often involve both repellents and attractants. I have been discussing the repellent aspects of the cave site. However, the site was more attractive than repellent: if it weren't, I doubt I would have trekked there in the first place. At the heart of this attraction is the cave, which is one of the most resonant symbols of prospect (and potential danger) known to us. That I was able to approach the cave is largely due to curiosity, without which our ancestors would have been shortchanged and hard pressed.

The idea that curiosity killed the cat is, then, something of a falsehood. Or perhaps it killed some cats, but certainly not all. In any case, there is clearly an adversarial relationship between curiosity and danger. The terrible desire we feel to explore and flee from what frightens us has become something of a staple for horror movies. The difference, however, is that in the movies the audience usually already knows what's waiting in the dark. So we protest by hurling insults, or we shrink in our seats, clench our teeth, and hope for the brave young man who instructs his weeping girlfriend to "Wait here" (as though "here" were somehow safer) as he investigates, only to turn up later as a hood ornament or as the local sheriff's breakfast sausage. We sit comfortably in our cushioned seats, thinking about what we would have done differently now that we know our protagonist is dead.

We would have never left our lovely mate–breasts heaving, muscles glistening–to fend for himself or herself. We would have gotten the hell out of there. Actually, on second thought, we wouldn't have driven out to the dark house or entered the dank cave in the first place! Ah, but therein lies the difference. For your average virile male, the first thought is to investigate. "Second thoughts are for old men!" cries our hero, pounding his chest. I must confess that although I do not possess the kick-ass-and-ask-questions-later mentality of our young protagonist, investigate is exactly what I did, despite my apprehensions.

Shortly after climbing the ledge, Greg went one way and I went another. I found myself standing alone near the entrance to the western-most cave in the series. I could hear Greg thrashing in the bushes and mumbling curses below me. I felt like I should wait for him before rounding the corner and entering the cave, but evidently he was going to be a while and patience isn't my strong point. The entrance to the cave is situated in such a way that I had to approach from the side, rather than from the front. This significantly minimized my ability to approach cautiously. Without the advantage of distance, I could not safely determine whether the cave posed any danger. Once I rounded the corner, I would be totally exposed to whatever might be inside. I recalled the run-in I had had with a javelina a couple of months ago and I wasn't thrilled about the possibility of being charged again.

I procrastinated by calling to Greg and asking him what the hell he was doing down there. Maybe my yelling would give the supposed cave dwellers an opportunity to flee. Greg's response was hard to make out, but I could tell he was still a ways off. I took a drink of water and made sure my five-inch folding knife was within easy reach. I considered drawing it, but then thought I might fall on the damn thing in the event I had to hightail it out of there. For a moment the descending song of a canyon wren high above me offered a reprieve from my apprehensions. I decided that the wren would not be singing if there were a dangerous animal in the area. Somewhat reassured, I returned to my preparations with renewed vigor.

I debated whether to unclip my binoculars from my chest but decided against it when I realized they might provide a barrier between my would-be attacker and me.

I was a bit annoyed for having left my hiking staff back at the ledge. I untied and retied my boots. As I stood up, I saw the top of Greg's bald and sweaty head emerge from the dense brush like some bizarre reenactment of a creation story. By then he was only a few yards away, and that was all the support I needed. As I rounded the corner, my heart pumped fiercely, and my tongue felt like it was wrapped in cotton. I couldn't see for a few seconds as my eyes adjusted to the darkness of the cave. My limbs surged with adrenaline. A euphoric, high-octane wave of blood washed over and awakened every cell in my body. I was ready for anything. My entire life had become about confronting this moment!

"Get your free nipple-piercing here!" Greg was a few feet behind me holding up his shirt and examining his chest. The cave trance was broken. "Don't pay to have your chesticles pierced," he said. "A palo verde tree will do it for free." I could see a small trickle of blood beneath his right nipple. It looked like a balloon that had escaped from an extremely diminutive person's hand. I tossed Greg the small first-aid kit and sat on a stone inside the cave. As far as caves go, this one was pretty small. But at twenty feet deep and twelve feet high, it was by far the most habitable of all the caves beneath the overhang, which were less like caves and more like pockets. The ground was lumpy with hoof prints except for where it had been smoothed by the sidling motion of a large snake—possibly a rattler. Someone had made a small fire in the back of the cave. The only other obvious human sign was a basket and a straining screen used for finding (and then stealing) artifacts. After the initial fear of rounding the corner, it felt good to relax in the cool of the cave. My risk-taking had paid off. I was still alive and even had a place to sit out of the sun. It wasn't much, but I relished it.

This moment-by-moment, linear account of my encounter

with the cave is somewhat deceptive, however. Writing is intrinsically linked to thinking and reflection, but that's not really what happens when we are forced to live in the moment. I wasn't thinking so much as I was being. Thus the process of deciding which precautions to take wasn't nearly as conscious as my description would suggest. That's not how an efficient response system works. Analysis and reflection may come after the fact, but during the experience I was more or less on autopilot. It's as though I had already completed the analysis and was benefiting from it at the time of the experience. Therefore, even though I had never personally encountered this exact situation before, in many ways I responded to it as though I had encountered it many times before.

The notion of being on autopilot helps to illustrate the nature of the Pleistocene mind. I was conscious of several things that should have allowed me to put down my guard, including the almost zero probability of encountering a dangerous animal inside the cave. Mountain lions aren't exactly plentiful in the Superstitions. In fact, the only large animal that would consider inhabiting this cave is the human animal. The likelihood of encountering a hostile human—even an artifacts thief—is similarly scant. And yet I acted as though I were expecting to be assailed. Ultimately, then, I was responding to conditions that haven't existed for thousands of years. Talk about being entrenched! In terms of affecting our behavior, the Pleistocene mind is always at work, but never more clearly so than in natural environments such as this. It is as though we begin the slow return to our true selves when in the wild. We meet ourselves there.

Greg had cleaned and dressed his wound and sat quietly in the pocket just east of the cave. I wondered how many people had sat in that exact spot and looked out across the desert and mountains. The more I looked around, the more I began to understand the appeal of this site. I had already addressed the issue of access. Greg and I had proven that we could reach the cave relatively easily by using the indirect route. But there was always the waterfall route, which would have been useful for quick descents to the desert floor. I also wonder if perhaps ropes were used to convey items up and down the waterfall. The cave is situated perfectly between the two hunting and gathering zones—the desert below and the mountains above. In any case, I no longer wondered whether the site was habitable. On the contrary, I could now see how, given the cave's southern exposure, the cooler months of the year could be spent here.

Lingering in the back of my mind, however, was the issue of the site's vulnerability and how the early inhabitants had come to decide that the site was safe, despite its apparent obstacles to escape. There's a wonderful scene in the *Quest for Fire* (1982) that may help with the answer. The film follows the adventures and misadventures of a group Neanderthals as they endeavor to learn and protect the secrets of fire. Over the course of the film the group encounters several bands of hostiles, all bent on plundering the group's booty, which includes women, weapons, food, and fire. Within the first ten minutes of the film, a violent battle takes place at their cave, deep in the mountainous woods. The key to the rival band's assault was that they were able to sneak in and attack from above the cave entrance. Once they had killed several of the group's members by hurling stones and logs down upon them, a second group of invaders attacked from the ground. Things did not go well for our Neanderthal protagonists who, in the wake of this two-pronged assault, barely escaped with their lives.

I saw *Quest* twenty-three years ago, but I still vividly remember the layout and the details of their cave. I might remember the cave so well because my Pleistocene mind was evaluating its inhabitability. In any case, humans living at that time and in that part of the world (what is now France or southern Europe) were like humans living anytime and anywhere: they made do with what they had. Although the site had only one significant drawback, it was deadly: the site did not offer views of the surrounding area and so could not easily be protected from a surprise attack. This would have been especially true with respect to the area above the cave, which, in addition to being precipitous, was heavily wooded. With all that cover, a stealthy predator or—as was the case for our group of cave dwellers—a hostile band of fellow humans would have had little trouble sneaking up on and dispatching a lookout. And once the lookout is gone, it's open season on the Neanderthals.

But therein lies the difference between the two sites. While the cave site that Greg and I hiked to has its fair share of difficulties, including an apparent shortage of escape routes, its occupants enjoy commanding views to the south, east, and west. In the context of vast desert stillness, the naked eye is very effective at detecting the slightest movement in the landscape. I remember seeing a couple of people on horses when I stopped to rest on the platform where I had found the pottery shards. They had to have been over a mile away, and still I picked them out of the landscape. I am sure my spotting effectiveness would have been even greater had I lived out here a couple thousand years ago: then I would have been looking for danger.

The location of the platform led me to believe that it was primarily used as a resting place for people on their way up the mountain. However, once I had visited the cave site and had gotten a better sense of its vulnerabilities, I realized the platform may have also been used as a kind of lookout tower. The pottery shards should have been the first clue that this was a place where people had spent their time. The tower sits atop a steep slope approximately halfway up the mountain, so clearly there was some resting going on there. Greg and I certainly took advantage of the area. But as the word "tower" implies, the platform is also extremely well-fortified and thus could only be accessed by ascending the steep south slope, which could have been easily defended. Approaching from the west—a high cliff face—would have been suicide, and approaching from the east and north is simply not possible given the lay of the land.

Similarly, compared to forest, the Sonoran Desert doesn't offer much cover for larger animals, a condition that would make it very difficult for a would-be attacker to go unnoticed by several sets of eyes that were looking for trouble. Not to mention that lookouts would have space and, consequently, time on their side. For the sooner trouble had been sighted, the more time lookouts would have to warn the others and to decide how to deal with the threat. In fact, I would argue that lookout towers or strategic vantage points would have been a condition for inhabiting this—or indeed any indiscreet—cave site.

When I have been out on the desert with friends and colleagues, I usually point to a randomly selected expanse of mountain and desert and ask them what they notice. I've asked over a dozen people. Not a Lou Harris poll, I admit, but out of the myriad possibilities, each and every one of my respondents zeroed in on the same things: mountain tops (prospect), riparian areas (prospect), unusual rock formations (way finding), and caves and anything that closely resembled a cave (refuge). Caves large enough to accommodate humans would have been especially important, mainly because there just aren't that many of them. And we all know what happens when there are too few resources to go around.

Cave C and the Neanderthal cave in *Quest* illustrate the point that caves pose different challenges to safety and therefore require their own unique brand of defensive vigilance. Such factors as cave size, location, number of inhabitants, and visibility would surely affect how inhabitants dealt with issues of defense. In terms of visibility, cave C is anything but discreet: as a landmark it can be seen from miles away. This is why lookouts would have been crucial to an effective defense.

Given the different means of access and defensive characteristics, as well as its location and exposure, the cave must have been a prized refuge.

In contrast to how we presently use and occupy our own homes, it seems that for early desert inhabitants no one place was all things at all times. This was probably true to an even greater extent for the Desert Archaic peoples, who traveled from site to site in order to meet their needs for refuge and resources, the type and availability of which would have changed over the course of the year. What was for them a dynamic lifestyle that kept them in touch with each other and with the land has, for most of us, become a relatively isolated existence. We drive from home to work and back again in our hunt for dollars. Instead of the earth we feel the gas pedal as we drive to the food, where we gather packaged meat and pick glistening vegetables from their pyramidal displays. Despite the myriad ways life has changed since the time of early desert inhabitants, these changes have more to do with the appearance of change than with actual change itself. Using the Desert Archaic peoples as a point of reference, over the last 9000 years humanity has seen many changes with respect to technology, but we—the apparent benefactors of current technology—have remained basically unchanged.

Sure, we may have gotten a little broader around the waist and face; and, paradoxically, we now live longer due to medical advances. But life is the same on the level of gene transmission. So although it may seem as though we bear no connection to early peoples or other animals, nor share their concerns or preferences for certain landscapes, in reality we are like one long river-in-the-making. On a fundamental level, our needs are the same as they were during the time of the Desert Archaic peoples and even further back in human and pre-human history. The only thing that has changed is the means whereby we attempt to fulfill our needs. But the Pleistocene mind is resourceful. It makes do.

In the end, I imagine my concerns about the cave's accessibility and vulnerability were very similar to the concerns of the cave's early inhabitants and that we went through a similar process of identifying challenges to access and safety and of determining how these issues might best be addressed. Before anything else, however, perhaps we shared an attraction to the cave, which prompted us to explore its potential as a possible refuge. That we would all be attracted to shelter must seem pretty obvious. How we actually decide whether a place is worth inhabiting—including the place's safety, proximity to resources, and the routes we take to get there—also seem like matters of common sense.

Consequently, this talk about biological or evolutionary predispositions for certain habitats might seem a bit much in the wake of something as ordinary as common sense. But exactly how does common sense become common? Whether we believe common sense developed in our own time or over the course of all time, one thing is clear: all animals, including humans, select habitats on the basis of how well they fulfill our basic biological needs. Given this fact and the universality of these preferences in humans (not to mention the existing research to the effect), the idea that culture alone can adequately explain the appeal, aversion, or ambivalence we feel for certain places is deeply unsatisfying to evolutionary biologists and their ilk.

That most of us feel a profound attachment to and take great comfort in our homes is undeniable. Our homes are our strongholds, our places of power, where we rest and sleep and take meals in safety. We rejuvenate and heal and plan. We feel joy at the thought of returning home after a long day of work and feel fear when we think of our home being invaded. We have a hard time appreciating what it means to be without refuge or shelter. But I don't think our indifference is necessarily malicious. I believe it is a problem of computation. Our individual survival is so intimately correlated with having a refuge or shelter that the phenomenon of homelessness usually doesn't register unless we experience it.

Of course, there are exceptions to everything I've said here, including the fact that many people take it upon themselves to advocate for the homeless. It is also true that the concept of home means different things to different animals. In comparison to how we define home, for instance, other animals must seem homeless. But that is the anthropocentric point of view. Narrow definitions such as this fail to recognize how similar our needs and decision-making processes are when it comes to selecting a place of refuge. The orb weaver's shelter may not have four walls and a roof, but it is a shelter that the orb weaver makes in order to protect herself from the elements and the watchful eyes of predators. Like us, the orb weaver is careful and choosy when deciding where to construct her web. She selects the site on the basis (or on the perception) of how well it will enable her to survive and reproduce.

People living in Tornado Alley may want to take notice: the orb weaver builds out of the wind. And like every human to have ever lived, she's attracted to places that promise an adequate supply of food, so that when the time comes to mate, she will have enough fat stores to get her through the lean days that precede and follow her reproduction. Thus, while the homes of different animals may not look the same, we each have evolved criteria for selecting them because, at some level, we have similar needs. If we think of needs as biologically driven incentives for behaving adaptively, then habitat theory is not just something that describes and connects us to early humans, but to all life forms that, for the sake of survival, have developed and passed on preferences for certain environmental features.

x. The Good Neighborhood

Location, location, location. Although this familiar expression usually appears in the context of commercial interests, the deep recognition that location is crucial to our personal and genetic survival is as old as our species. In fact, early humans usually selected their shelters on the basis of where they were located, for without access to prospects such as food and water, "shelter" would mean precious little. Given our ever-increasing dependence on technology, our mobility and the resulting movement away from the lives and challenges of early peoples, it may be difficult to appreciate the extent to which we retain this ancient sensitivity to location. But clearly our concept of neighborhood reflects our own primal and often unconscious need to select the most optimal environment in which to live and, ultimately, raise offspring.

So what are the components of a "good neighborhood" and why might our attraction to such places have been and continue to be significant? Gordon Orians and J. H. Heerwagen provide us with a useful starting place by noting that we make high-and-low level decisions when determining how to use and respond to habitats. "In general," they argue, "the time and energy invested in a decision should be positively correlated with the importance of that decision to the fitness of the individual." When Kim and I first began our search for a home, we must have looked at fifty different houses in half as many neighborhoods before we finally found our home in Gilbert (we went through the same process a few years later when we moved back to Utah). The process took roughly four months. True to predictions made by Orians and Heerwagen, habitat selection is generally a very long and involved process—and wisely so. Choosing hastily or poorly could have serious consequences.

Given that most if not all humans spend considerable time

selecting a habitat, it seems that this tendency is an example of a biologically prepared learning bias. Animals that choose carefully (like my little orb weaver) and with certain criteria in mind, stand a better chance of surviving and passing on their genes than do animals who choose hastily or who, because of unfavorable social conditions, are forced into "bad neighborhoods" or, to use another familiar expression, onto the wrong side of the tracks. Ecologists call this phenomenon "environmental racism"—people of low socioeconomic status are often forced to live in environmentally degraded places. The reasons for this become apparent when we consider their alternatives, or the characteristics that make a neighborhood "good."

As the work of Orians, Appleton, and others shows, landscape paintings offer valuable insights into our habitat preferences and aversions. I like to watch *The Joy of Painting* with the late Bob Ross every Saturday morning. Ross was almost exclusively a landscape painter, so his work represents an ample source of habitat data. I recall an episode when Ross was painting a scene deep in the forest. As he painted several large, dense pines in the background of the painting, he said that it was "important to have these trees in the distance." With the trees in the distance, the painting's foreground is open, welcoming the viewer into the painting, and, once he is there, offers easy access and mobility.

The majority of Ross's forest paintings are strikingly similar in this regard, and even when the viewer is inside the forest, rather than on its periphery, there is usually some form of refuge (such as a small cabin), and the trees tend to be fairly well spaced, filling the forest with equal quantities of light and shadow. The presence of light not only enables the viewer to detect any dangers that might lurk in what Dante called the "dark wood of error," but facilitates movement and the identification of ways through, and perhaps out of, the forest.

Our typically anxious associations with dead ends reflects the Pleistocene preference for multiple routes. One-way-in-

and-one-way-out just doesn't sit well with most of us, which may help explain why so few housing communities feature them. If, like access to schools and shopping or mountain views, number of escape routes were used as an explicit selling point, our house here in Gilbert would be highly desirable. Our house is situated on a "T," which means I can walk out my front door and get out of the neighborhood by going straight, turning left, or by turning right. This aspect of our habitat selection reminds me of Cave C and its multiple access and departure routes.

Despite our own increased mobility, Kim and I were determined to reside near our places of employment. Normally we describe this sort of decision making as a matter of convenience. But it also reflects our automatic application of cost-benefit analysis when selecting a neighborhood. If in the process of obtaining food we burn more calories than the food itself provides, we would suffer a nutritive deficit and would slowly starve to death. Clearly proximity to resources is an important facet of habitat selection, although it is sometimes difficult for us to appreciate this need because we are typically inundated with resources. Like most people in suburban areas, Kim and I live within two miles of at least four different grocery stores. But for people living in "undeveloped" areas, the distance may be considerably greater. Kim's parents live in a newer housing development about seven miles southeast of us, and up until about a year ago—which is when the first grocery store was built in their immediate area (actually, on the same block!)—it was not unusual to hear them complain about having to travel so far (four miles) to get groceries.

For many of us, then, living near the workplace and reliable resources is simply a matter of—that's right—common sense. If it is so important for us to live near our work, it seems odd that more people don't live in urban/industrial areas. There are many reasons for this, including lack of affordability, but beneath those reasons is something I will refer to as the "waterhole" phenomenon. I've already talked about riparian areas and why—to borrow from Frost—they would have made good neighbors. As sustained sources of water, waterholes make excellent neighbors (kind of like the neighbor with the only swimming pool on the block). Living in Arizona, I take special interest in waterholes whenever I encounter them in the wild. So too do many other species whose signs—tracks, scats, feathers, hair—are everywhere. Consequently, waterholes can be very dangerous places, which is why most animals drink quickly and then head for the hills.

Similarly, our places of employment are usually in populated areas. Although they are not waterholes, they are characterized by higher population density and human activity, as are grocery stores and shopping malls. Where there are more humans—especially unfamiliar or strange humans—there is the potential for more danger, either because of the humans themselves, or because of other animals and microorganisms that prey on humans. Maybe this has something to do with why men (young to middle-aged, perhaps?) are notoriously anxious and unhappy shoppers. (Kim leaves me at home.) Surely many men are averse to the activity itself (men are typically the hunters, not the gatherers), but it is also possible that certain-aged men get anxious when shopping because their Pleistocene minds perceive grocery stores and malls as places of danger. It would be much easier to stay at home, watch football and thus confront these dangers vicariously through the blood and sweat of other men. For what fanged terrors lurk down by the meat case? And what about that man lingering in the beer aisle with a bag of Doritos and a packet of weenies under his arm? Is he looking at me? My mate?! My *child*?! I'll rip his throat—oh my, but isn't she lovely...sans the guy with his arm around her waist. Ah, how the primal drama unfolds each and every day of our lives. And I thought all I was doing was buying diapers and a newspaper.

Obviously this is an internal monologue—these thoughts

seldom find physical expression. But they are there if needed. Our ability to check our Pleistocene predispositions and inclinations—however pronounced or latent—illustrates the yin and yang, as it were, of our evolutionary development. We may have certain thoughts or feelings, but it is not always wise to act on them. Although examples to the contrary abound, self-control is a crucial complement to our various survival instincts. Without it, society as we know it would not exist. Nor would men take care of the children, or women win the bread. However, it seems that some situations are so charged with survival relevance that they continue to elicit predictable responses.

We like the idea of living near the waterhole, but not so near that we are constantly subjected to the dangers associated with high traffic areas. As a rule, then, we visit the waterhole only when we must. Many people live in the cities and seem to manage as well as any suburbanite. Perhaps one of the ways city dwellers manage to live among so many other humans is by living as high up as possible. Some urbanites are so fortunate as to live above their workplace. This must be the urban fantasy, the suburban equivalent to working at home. Not everyone can afford to live in a penthouse with a view of Central Park, however. The majority of city dwellers live in considerably more humble dwellings. So how do they even the odds and compensate for what is undeniably a stressful and often dangerous environment? Why do people select such apparently inhospitable habitats?

I'm sure city dwellers each have their reasons, but I wonder if, generally speaking, those reasons might not all be listed under *resource availability*. Unless we make the mistake of assuming that "resources" means "champagne, caviar, and private schools," we can see how the perception of resource availability is relative. This is not to deny, excuse or justify the social inequities that prevent many people from having a roof over their heads, let alone from having a dish of caviar.

Instead, I am suggesting that humans have an innate attraction to fecund habitats, and that the element of risk associated with obtaining resources is unavoidable. How, if, and why some humans acquire those resources or not are questions of proximate relevance. Any number of factors, including one's socioeconomic status, age, and physiological state affects the likelihood that resources will be acquired, but our attraction to places that promise resource abundance might very well be a universal among animals.

Perhaps a useful synonym for "resource abundance" is "opportunity." Compared to the opportunities available to city dwellers, suburban life must seem like a wasteland. But from the point of view of attraction, which varies from person to person, the comparison is moot. Kim and I were concerned with selecting a neighborhood in proximity to opportunities. Parks, quiet nights, and cleaner air are not opera houses, fine dining, and art galleries, but they still represent opportunities or prospects. This underlying attraction to prospects has to do with why the majority of people live in such high concentrations. In addition to environmental cues, the presence of humans is itself an indicator of resource availability. New York City is an example. But strange humans can be dangerous, especially when there are so many of them. So it seems we are in a bit of a pinch. After all, dead people don't eat ice cream, buy and sell stocks, listen to Mozart, conceive children, or anything else for that matter.

The benefits of living in a sea of people have got to outweigh the costs. Otherwise it just wouldn't make sense to live so closely to other humans. Regardless of where they live—city, suburb, forest, mountain, desert, ocean—humans and other animals are attracted to habitats that balance their needs for prospects and safety. Seems simple, but how is it that affordance-rich habitats, which often attract high numbers of varying species, retain some measure of safety? Put another way, how is it that Kim and I came to reside here in Gilbert, along with a few hundred-thousand other inhabitants, a significant percentage of which does not have our well-being in mind?

Dangerous humans are a fact of life. They always have been. Sociobiology, which attempts to explain social behavior using evolutionary theory, suggests that every one of us behaves in ways aimed at reducing the probability of being victimized. We have developed multiple strategies for reducing the likelihood that we will end up at the business end of someone's ill intentions. One strategy is known as the dilution effect. Kim and I have observed this strategy in action while fly fishing for trout in the White Mountains of eastern Arizona. Mayflies hatch by the tens-of-thousands on cool summer evenings, usually after a rainstorm. During these times, the lake boils with trout as they chase the emerging insects. The flies are encased in a balloon of air that hastens them toward the surface, where they break into flight. By overwhelming the trout's consumption capacity, mayflies increase their individual chances of survival. Of course, humans are not mayflies, but it is possible that one reason for why we live in such high concentrations is because we also benefit from dilution effect.

Perhaps the related but sugarcoated strategy of safety in numbers is more palatable from a human perspective. Many of us don't respond well to the idea that we aggregate simply to reduce the chance that we'll get eaten. Incidentally, there is a term for this aspect of dilution effect—the appropriately named "selfish herd," whereby some members of the group are used by other members as shields against predation. Selfish herd makes the dilution effect seem innocuous by comparison. Given the fact that members inhabiting the periphery of the group are more likely to suffer attack and predation, one wonders if there is a similarly high probability that people living on the peripheries of their housing communities or villages will be the victims of crime. This may also help to explain why many of us resist living in houses that border major roads, which may be used by "predators" as access points and escape routes. Thus, selfish herd theory predicts that we will generally favor living in houses that—all else being equal—are closest to the center of the community.

This notion of community points to additional strategies for dealing with danger and, consequently, for reaping the advantages of living near other humans. Kim and I have lived in Gilbert for about three years now, and over that time I've gotten to know some of my neighbors. However, I've made no effort to know the neighbors other than those on either side of my house and directly in front of it, nor have they made any effort to get to know me. I realize this has partly to do with access, but I think the more convincing reason is that my closest neighbors are in the best position to look out for my interests. In return, I look out for theirs. Living so close to and maintaining peaceful relations with other people in the community have always required considerable tact. Sociality is, then, surely an adaptive strategy, particularly as humans formed larger and larger groups of unrelated people and increased the potential for conflicts. I think this is why I instinctively limit my interactions with all my neighbors, and why my neighbors generally do the same. Unlike early humans, we tend to stay in the same place for years instead of weeks or months. But we all know that spending a day in conflict with another person is extremely taxing, so it makes sense to keep some distance between ourselves and other members of our community.

If my hypothesis is at all correct and we instinctively limit our interactions with community members in order to avoid conflict, then we can predict that whatever interactions we do have will be significant. In fact, it could be argued that even the idlest banter is fitness enhancing, to the extent it reaffirms the relationship of participants. I first became aware of this possibility while hiking the Gold Canyon Trail in the Superstition Mountains. The trail is popular with locals and out-of-towners, so it is not uncommon to encounter other hikers. Generally speaking, when it comes to strangers I am friendly but cautious, and I think most other people are as well. I think this is why I noticed how peculiar my exchanges were with strangers when out on the trail.

I hiked the Gold Canyon Trail five or six times over a period of two months and during that time I encountered twenty-five or thirty people, some of whom did little more than smile or wave "hello" as we passed each other on our way into or out of the desert. And still another percentage didn't even offer that much, nor even break stride. Instead they averted their glance and kept hiking. But the majority of people was very enthusiastic about chatting and would often stop for the purpose. I find their willingness to stop and chat remarkable for a few reasons. The first is that as a rule we do not normally engage strangers, and if we do, we usually do not go beyond acknowledging them with a "hello" or a quick smile. I also noticed that hikers who were on their *way into* the desert were more likely to initiate conversation than were hikers *returning* from it. Interestingly, this dynamic does not seem to occur under any other circumstances. We do not usually stop people on their way out of the grocery store or shopping mall and ask them questions: "Did you see any cabbages?" or "Any sales I need to be aware of?"

Out in the wild places, something special happens that inspires us to put down our guard when it comes to engaging strangers. After all, we are not going into an air-conditioned shopping mall; we are returning to the primal space inside each of us. Because we are going into a real and potentially dangerous environment, we look to increase our chances of avoiding problems, which includes gleaning survival-relevant information from fellow travelers. When I have either participated in or observed an exchange between strangers on the trail, hikers were without exception interested in learning all they could about what lay ahead. Whereabouts of water, wildlife sightings, directions, and clarifying distances from and time needed to reach their destinations were all topics of interest.

Humans who, for whatever reason, did not seek survival-relevant information in the ancestral environment would have been exposed to all sorts of preventable dangers, such as the presence of dangerous animals. Thus, despite our reservations about one another, being social has its advantages. Nowhere is this more apparent than in the wilderness, where we are open to the elements and to other dangers. When compared to these wilderness exchanges, the survival value of interacting with our neighbors may seem negligible. But the usefulness of sociality extends to every situation where survival is an issue and dangers are present. Just the other day, for instance, Kim, Wilder, and I were out on our morning walk through the neighborhood when we encountered a woman and her dog. As we neared them, I could see that the woman was upset. Her cheeks were flushed and both she and her dog were breathing heavily. Before she had even reached us, she blurted that bees had just attacked her and her dog. She avoided injury, but her dog had been stung several times. We spent the next few minutes clarifying the bees' location, discussing the threat they posed, how the problem should be addressed, and so on. We even shared stories of past encounters with bees! Gone were all the precautions for dealing with strangers. And yet we didn't even exchange names.

Africanized bees (also known as "killer bees") are very dangerous, and escaping them would have been difficult even had I been alone. But with a twenty-pound toddler strapped to my chest, chances are things would not have gone well at all. Kim and I were deeply grateful to learn of the bees' whereabouts so we could adjust our route. Clearly *we* had benefited from this stranger's information, but what had she gained? Unless one believes in altruism, sociality must benefit both the information donor and the recipient. Perhaps one way we might answer this question is to ask what the woman needed subsequent to her frightening encounter. She had not been physically harmed, but she was psychologically distressed, in which case Kim and I did our best to comfort and console her. She needed our sympathy, and that is what we gave her in exchange for helpful information.

By offering the woman our sympathy, Kim and I helped to fulfill her immediate needs, but there are long-term benefits involved as well. In many ways, housing communities or suburbs are just modernized villages. Both are comprised of individuals who share the same environment and who can therefore expect to encounter one another from time to time. In fact, Kim and I often see the woman on our daily walks. Given our history, we are cordial, which is important, but what is more important is that should the woman ever require our aid, Kim and I will be more inclined to give it to her as payment for alerting us to the bees. Reciprocal altruism, then, is salient wherever groups of people co-exist, especially unrelated humans who otherwise would have no incentive for watching out for each other.

As we saw when considering peoples' responses to certain landscapes, adaptive behavior is relative to the individual and depends on several factors. This helps me understand how there can be so many strategies for dealing with the same challenge—in this case, of living among strangers. It also seems that people whose situations are similar will employ similar strategies for dealing with community life, maximizing the benefits while minimizing the costs. Although my immediate neighborhood is comprised of families and two single men, I have invested the majority of my energy in getting to know the other families, especially those who, like Kim and me, have small children. Not only do we share space, but our familial concerns—among others—are similar. Perhaps more to the point, we face the same threats.

I wonder if this isn't at least partly why the single men who live around me tend to keep to themselves. Different species of social animals have various ways of responding to single males within their communities, but clearly some animal societies are more accepting than others when it comes to tolerating lone males. For example, when male coatis reach sexual maturity they leave the group and go off to form their own territories. Wild mustangs use a similar strategy, but males typically band together for a time before seeking out females, which, incidentally, sometimes "belong" to mature males. We may be much more tolerant of single males than the majority of other animals. Still, I don't think I've spoken more than a paragraph of words to the single man living across the street, nor has he said much to me.

Does that mean that I'm not looking out for him and that he is not looking out for me? Not exactly. We both benefit by default, or by virtue of sharing the same habitat. We may not talk to each other, but we at least watch out for each other's property. And whatever is threatening him could cross the street and threaten me. It makes sense for each and every member of the neighborhood to keep an eye on things. Whether we are single or not, keeping to ourselves is a perfectly viable strategy for dealing with strange humans. But when it comes to reciprocating favors, we usually get what we give—because, in general, we give to other humans who share our circumstances. Connecting with my neighbors is just one more way to augment my family's safety and garner resources, mostly in the form of valuable information.

Although the specific elements of "valuable information" may change over time and circumstances, in the ways that are most important to us, the definition remains the same: information is valuable (and rewarded) to the extent it enhances fitness. Knowing the whereabouts of killer bees is fitness-enhancing. Perhaps the most important reason for interacting with like-minded neighbors is because they may have information about potentially dangerous people living within the community. Due to competition for habitat, mates, and resources, humans (particularly men) have always posed the greatest danger to other humans. Consequently, providing and acquiring information on their whereabouts and intentions remains very important. I often reenact this primal exchange with Iliya, my neighbor across the street. Iliya was born in Nigeria and came to the Midwestern United States when he was a teenager, where he lived with a Christian family and became a devout Christian himself. He refuses to tell me when he was born, but he couldn't be any older than me, which puts him somewhere around forty. For a fairly young man, he is quite wise in the ways of people and the Earth, a fact that I attribute to his Nigerian upbringing.

In the evenings, I like get out of the cave that is my air-conditioned house and reconnect with the outside world. In the summer, I may wistfully check the sky for storm. I often find Iliya busy cutting his grass or caring for his trees. There is something timeless about humans standing beneath the trees, and so I go and join him. Mostly we talk about how best to care for our yards and other related matters. Iliya is very knowledgeable about desert-adapted flora and irrigation. He also knows how to build and maintain a house, and on a couple occasions, he has either advised me or helped me repair my dwelling. I consider him a great resource. In the past I have edited and proofread his writing, so we use a kind of loose barter system.

Another dimension to our relationship is gossip. For the record, men rival women in their enthusiasm for gossip. As with other types of information exchange, there is a donor and a recipient. Usually, however, Iliya gossips and I listen. He's lived in our neighborhood for about eight years longer than me, so he has the corresponding repository of information. In light of everything I've said about the "you-scratch-my-back-and-I'll scratch-yours" philosophy of reciprocity, this may not seem entirely fair. But Iliya does benefit by sharing the neighborhood dirt.

Last week Iliya and I were having our weekly discussion beneath my jacaranda tree (sometimes we will stand in the street like two lions affirming our territory, but typically we alternate between our yards) and the conversation turned to the neighbor just east of my house. Dino had just sold his house and was preparing to move. I told Iliya that I hoped the new neighbor would be as agreeable as Dino had been and he just smiled. I sensed Iliya didn't share my feelings and facetiously asked him if he were sad to see Dino go. Initially Iliya was reluctant to talk about it, but after some gentle prodding his tongue loosened and he delivered a sermon-style tirade. Apparently Dino had been showing a little too much interest in Iliya's wife, Brenda, when Iliya was away at work. According to Iliya, Dino attempted to woo Brenda by expressing interest in her and Iliya's three children. I wondered if perhaps Iliya wasn't just being paranoid, so I asked him if this had occurred on several occasions or if it had happened only once or twice. Iliya lowered his voice and said that this had happened on several occasions and that the way into a man's house is through him, not through his wife and children. "Indeed," I said.

Iliya went on to explain how in his culture, a man simply did not talk to another man's wife when he wasn't around—that is, assuming he talked to her at all. A nanosecond later I had reviewed my memory and could not think of a time when I had talked to Brenda sans Iliya. I then felt free to make the point that most cultures discourage adultery and extramarital affairs. Iliya has a degree in Environmental Resources, so I don't hesitate to incorporate biological hypotheses into our conversations. "Isn't it interesting how our cultural practices, however different, parallel the evolved function of male jealousy and suspicion?" Iliya may think of himself as a devout Christian, and I may think of myself as a dedicated, albeit burgeoning naturalist, but above all else we are *males*. If we even *imagine* another male variously cuckolding our mates, wives, and mothers of our children, we experience considerable emotional distress.

As mates and parents, men and women invest a great deal in each other and in their offspring, so it makes sense that we would be suspicious of other males and females who could possibly commandeer our interests. For the male, this could mean a lot of things, but your average Joe is deeply resistant to the possibility of providing resources for the mailman's offspring. He also fears losing his mate outright, an event which, if he doesn't already have children, would put an end to his chances of having offspring with that woman. And then he thinks of all the other women he might have wooed in the time it took to attract and lose this one!

Assuming our hypothetical male does have children, deep in the back of his Pleistocene mind he knows that if he is replaced there is a possibility his offspring will suffer neglect, abuse and, in some cases, violence at the hands of the surrogate parent. The majority of non-biologically related male parents do not succumb to the socially undesirable and selfish impulses of the Pleistocene mind. This is not to say, however, that women do not already repel the amorous attentions of other males with as much fervor as they use to fend off other females who may be interested in their mates.

Thus, common sense mirrors evolutionary psychology to the extent that both sexes know that the biological parent will generally do the best job of caring and providing resources for their shared offspring. Women tend to be the primary caregivers of children, so we would expect them to have evolved behaviors that facilitate, as opposed to undermine, that tendency. This may help to explain why, on average, women are much less likely to have extramarital affairs than are men, who may run off and increase their direct fitness by having multiple children with multiple wives. In either case, with so much at stake we should expect that any encroachment would be met with a swift response.

Iliya responded by giving Dino the cold shoulder, and when that didn't work (the Pleistocene mind is determined, but as a rule it is conservative with respect to energy expenditure and risk-taking), he flat out told him not to come around the house anymore, period. Suffice it to say that that was the end of their neighborly relationship. No wonder Dino decided to move! I am sure there were other reasons (apparently Dino wanted to live closer to the golf course), but given the seriousness of his encroachment, I wouldn't be surprised if his confrontation with Iliya was the primary reason.

By putting the moves on Brenda—Iliya's wife, mate, and mother of his children—Dino undermined his place in the community. Although he had proximate reasons for moving, perhaps his Pleistocene mind recognized he could no longer reap the benefits of living in our community. He had crossed the line, and when a villager crosses the line, other villagers talk about it. In effect, Dino crossed everyone who plays by the community's rules. These rules are articulated and enforced through cultural practices, but they stem from and are tailored around the biological challenges we face as social animals.

Iliya and I have at least two important things in common: we are each married to a young woman well within childbearing age and we have small children. Given these shared concerns, a threat to him is a potential threat to me. Although I have never doubted Kim's commitment to me, and feel no threat from Dino, if there is any male in the neighborhood in search of fertile young women, the Pleistocene mind wants to know about it. Clearly I stand to benefit from Iliya's information sharing, but what does he gain by putting me in the know? There are multiple benefits to sharing such information. For one thing, Iliya enlists my support in response to the current threat and in detecting future threats, which is a benefit he would reciprocate should it become necessary. The tacit agreement seems to be this: "I have given you valuable information that may help you to protect your interests. In return, I ask that we work together to address future threats."

This method of I'll-watch-your-mate-if-you'll-watch-mine could have been a standard benefit to groups of humans. It also seems that the various cultural attitudes toward and expectations of women ultimately stem from evolved male jealousy. Some cultures expend considerable energy in concealing women from view, which they accomplish through strict dress codes, and by keeping them indoors and out of sight as much as possible. Even more troubling is that men in these cultures sometimes commit acts of grotesque barbarism toward women who are alleged to have transgressed these expectations. In contrast to my wife Kim, who spends most of her time inside or in the backyard, Brenda socializes in her driveway with her children and the next-door neighbors. This is apparently what made it possible for Dino to make his move. Kim is seldom out front, and rarely says more than a few words to anyone, least of all to a man who clearly has more on his mind than the weather.

However unique a culture may appear, each shares the same impetus of human sexuality and other socially problematic aspects of human nature. Perhaps culture is the means whereby different groups of humans encourage some aspects of our evolutionary past while discouraging others, depending on how well these aspects advance their own interests. The attempts of some cultures to mandate chastity, for example, or to keep women out of sight, reflect the lengths to which males will go to guard their genetic interests. Interesting to think our own culture—despite its apparent sophistication—might have emerged to address the primal tendencies of the Pleistocene mind.

xi. Yards, Death, and the Healing Sleep

Kim and I were not unlike early humans and other animals when selecting a habitat. That is, the most we could hope to do is choose an area that balanced our needs for prospects and safety. Obviously, compared to humans living even as recently as 1,000 years ago, modern humans play a decidedly greater role in ensuring prospects and creating safe environments in which to live. Given our ability to change the environment on such a massive and often catastrophic scale, and that we typically move into "prepared" habitats or neighborhoods, it may be hard for us to imagine what it meant and still means to have a minimal effect on the environment. Compared to us, our early ancestors did not significantly impact the environment at large, but they did manipulate the areas around their dwellings, clearing them of debris that may undermine visibility or conceal a potential danger such as a snake or predator. Perhaps they encouraged certain plants and vegetation while discouraging flora that had no nutritive, medicinal, or protective function.

Insofar as we invest hundreds, if not thousands, of dollars creating and caring for our yards, Kim and I are like the majority of people in our neighborhood. We enjoy cultivating lush green lawns, gardens, and various species of plants and trees, but typically we do not depend on them for our survival. We still construct elaborate habitats, replete with flora, light, and water features—all characteristics that would have been advantageous in an ancestral setting. This suggests that our interest in constructing habitats persists because it was and continues to be adaptive. If so, then our preferences for certain landscape features—for certain kinds of yards—may offer valuable insights into the deep history of our species.

When my son Wilder was not even a month old, we had made a ritual of going outside each morning and checking on the yard. The notion of "checking" strikes me as slightly amusing—but over the course of a night, more than I will ever know can happen in a suburban yard. I am quick to note even the most subtle changes: cat tracks in the flowerbed, a dove's feather in the grass, a new spider web in the orange tree, a shed hibiscus flower on the walkway, a freshly dug lizard burrow. "Checking" may be an accurate description of what we do and, I would argue, of what we have been doing for millennia. The question is, why? Assuming Wilder and I are not the only ones to wander in their yard, why might humans have developed this behavior?

Seven months later, as I walked through the yard with Wilder in my arms, the ritual had become more elaborate. I shared the names of things; grass, dirt, rocks, cloud, sky. I tried to alert Wilder to a cicada clinging to the corner of the house, but he concentrated on the bigger picture. We strolled to the other side of the yard and stood beneath my favorite tree, the Chinese elm. I described the tree part-by-part. I began with the crown, then worked down through the branches, leaves, and finally to the bark and roots. I wanted him to know that not everything real is readily visible. That sometimes we've got to search. A cicada began to buzz, reaching a full trill. Wilder flapped his arms in excitement. At eight months old, his concept of gravity was merely a glimmer. He would crawl off the edge of a cliff if I let him. Still I noted that he was silent and did not wiggle in my arms.

I think back two hundred thousand years ago and watch a female hominid collect food from the ground while holding her infant. Weary of the tall grass, she prefers to travel across the open savanna. She makes her way over the rocks and through a grove of scattered trees, each with a broad canopy that provides shelter from the searing sun. The child is still and silent. One wrong move and he could slip from his mother's arms. An outburst could alert a nearby predator. When I look at Wilder, I wonder if these ancient children are the reason children to this day remain still and mostly silent when in their parents' arms. Granted, my three-year-old nephew Steve wiggles like a just-caught fish. He is spindly and well-muscled. But at three-and-a-half feet tall, he is much closer to the ground when I hold him. Still, even he doesn't attempt to break free all at once. Instead, he wriggles through my arms and slides down my body as if it were the trunk of a tree.

The early morning sun fell through the branches and patterned Wilder's face. I waited for him to look at me and then

I likened the bark to human skin. I explained how it protects the tree in much the way our skin protects us. He leaned and reached toward the ground, and down we went. Wilder was especially aware of larger objects, but he did not miss one of our box turtles as it scuttled across the flagstone on its way to water. Wilder's eyes widened and he grunted. The sound is vaguely simian. Kim thought this was his way of asking for more information. "Turtle," I said. He crinkled his nose and laughed. A bird called from the neighbor's tree and startled him into the moment. He looked at me for reassurance. "Is there danger?" his face asked. We couldn't see the bird, but I gave him the name anyway. "Grackle." The grackle called again and Wilder seemed less surprised. The day was a litany of such rituals.

We've been doing this nearly every morning for the last eight months, just before we go on our morning walk through the neighborhood. By the time Wilder was two years old, he had a pretty good sense of his immediate environment and recognized many of its inhabitants. But he also had a nascent knowledge of the much larger area of Settler's Point. The timing couldn't have been better: most children start walking by one year, gaining greater access to the environment. Most three-year-olds generally do not wander from their mother's side let alone away from home, but familiarizing children early with the ins-and-outs and dos-and-don'ts of their habitat makes sense from the point of view of prevention and safety.

Children have their own defense mechanisms, but these are no substitute for parental care. The more that parents offset potential dangers, the greater chance their offspring will carry on the genetic legacy. Perhaps this helps explain why each morning Kim, Wilder, and I see so many other parents walking with their children through the neighborhood. The importance of neighborhood walks depends on one's station in life, but surely one reason *parents* take walks with their children is to familiarize them with the layout, dangers, and prospects of their surroundings. Most parents have no idea that their walks serve this function. Therein lies the beauty of evolution. As the recipients of hundreds of thousands of years of genetic trial and error, most of us are natural born survivalists. Natural selection has acted on every cell in our bodies, all for the sake of genetic transmittance. And just as our bodies perform hundreds of functions each moment without our knowing it, so too do we act in ways that enhance our survival—whether we recognize it or not.

If it's true that we generally behave in ways that promote survival, then the entire spectrum of human experience and behavior could be regarded as data. We can ask questions about the adaptive significance of everything from the emotions to sexuality to social behavior to parental investment and so on. I suppose I like asking questions about my yard because that environment is so accessible. I walk outside my door and I'm literally *in* my subject matter. I'm beginning to understand how biologists like my colleague John Alcock can spend entire mornings walking a ridgeline in pursuit of a certain species of wasp.

Biologists, naturalists, and children are students of the real. The rest of us seem content to live our lives in the self-stylized realm of our own subjectivity. We can't deny the importance of this realm (it exists, after all), but this does not mean we cannot inquire about the world beneath our feet and discover what might well be universal among humans and other animals. Researchers have documented the restorative properties of natural environments (the hospital where Wilder was born features dozens of landscape paintings and an indoor arboretum with a waterfall). They have proven what we all know—whether we stand in an old growth forest, on a beach, or in our own backyard: nothing stirs, frightens, disturbs, pleases, rejuvenates, and restores us like our encounters with the natural world.

I know some of these speculations might seem a bit "out there" given the utter familiarity of yard care and daily walks and the other mundane activities I've discussed. There have got to be more simple explanations for our behaviors. We might create verdant yards filled with diverse flora because it eases tension, and take walks just to get out of the house—but notice how neither of these explanations conflicts with an adaptive hypothesis. Notice, also, how neither explanation tells us much about why we do the things we do and feel the way we feel. Why does creating a verdant environment ease tension and produce pleasure? As a rule, humans prefer to keep things simple, so it is understandable that we are often all too happy to forego involved explanations in favor of what is immediately proffered. When it comes to the explanation menu, most of us prefer "the usual."

The truth is, however, that every facet of human life might be better understood by examining it in the context of our evolutionary past. In the end, and despite our intelligence, we really have to try to live up to our self-congratulatory view of ourselves, and work at thinking and acting in ways that help us better understand and, when necessary, live contrary to our evolutionary heritage. For without a more complete explanation, we cannot hope combat the unfortunate aspects of our evolved responses. Making war and killing other humans is part of our evolved nature, but together with the destruction of the environment, these constitute the greatest preventable failures of our species.

Early in *Jeremiah Johnson*, Johnson tracks the Crow warriors who murdered his family. He finds them in a grove of birch trees. It is late spring and the sun is low on the horizon. The Crows wear smoke-gray skins and dark war paint. A rabbit roasts in the fire. The Crows sitting near the fire resemble the ground's alternating patches of earth and snow. Those who stand are practically invisible against the background of trees. Flesh, earth, tree, sky, snow, blood, hunger, and revenge commingle in the pale afternoon light. Then Johnson fires his long-arm and the killing frenzy begins. Within moments, the last Crow flees the camp. Johnson pursues him into a clearing, where the Crow slips and falls in the snow. Armed with only a knife, the Crow turns toward Johnson, kneels in the snow, and sings.

Until now, we have not really seen the faces of the Crow. Redford knew that faces belong to individuals, and until this moment he wanted to show us how easy it is to destroy the faceless masses. Soaked in adrenaline, Johnson breathes heavily. He has just killed a half-dozen men. Their blood mixes the way blood will. The Crow sings. At first, I think his singing comes too late. I am caught in Johnson's bloodlust. But the beauty of the Crow's song gives me pause and awakens me from my murderous trance. Suddenly, killing this man seems pointless. Johnson lowers his knife and walks back into the sheltering trees. He curls into a ball and sleeps a long, healing sleep. It is not too late. We could follow him.

IMPLICATIONS OF AN ECOCHILDHOOD

Caribou, Maine 1968 – 1979

i. I Killed; Therefore I Am

Kim, Wilder, and I stood outside in the yard and watched as the first monsoon of the season rolled in from the south. The storm was a remnant of a hurricane that made landfall in Mexico a couple of days ago, but it was still powerful enough to generate a wall of dust thousands of feet high. Behind the wall, a rain-cooled, sixty-miles-per-hour wind brought relief and destruction. We ducked inside until the dust had passed. The neighbor's palm thrashed and bougainvillea leaves swirled into the air. Eager to enjoy the sudden twenty-degree drop in temperature, we stepped outside and breathed the sweet, earthy smell of rain. Apparently our neighbors shared our enthusiasm: I could hear them in their backyard, laughing and commenting on the color of the sky, now black as deep water. As the storm approached, we retreated to the covered patio and waited for the spectacle to commence.

This was Wilder's first monsoon—but Kim and I were so excited, it could have been ours, too. We could hear the rumblings of thunder and the far-off wail of sirens. The doves appeared black against the gray-brown sky and seemed to fly wildly and with great pleasure as the wind burst through the trees and over the roofs of houses. The rain fell in huge, cold drops. Kim and I could not resist, and so together with Wilder we stepped into the rain and stayed there until our hair dripped and our bodies shined beneath a skin of water. Wilder laughed a deep belly laugh and we all laughed and shivered and forgot the freakish heat of late July.

One of the things I cherish about Wilder is the same thing I cherish about my closest friends: a capacity for wonder and curiosity about the everyday world and the simple things like rain. Wilder has a lot of choices in playthings, but still he selects only a fraction of the toys available to him. Because he plays with only a handful, we might think he'd choose toys with all the "bells and whistles." Interestingly, just the opposite is true. Wilder's favorite playthings are rocks, dirt, leaves, grass, water, and now rain. As he gets older his interest in complexity will increase, but at this stage in his development, he seems happy to pass a rock from hand to hand, pull grass, tear leaves, and clutch dirt. The rain must be a fascinating novelty. The big drops drench him and run into his eyes, and judging from his amazed look and his laughter, I think he must—like Kim and me—feel something like rapture. Rain is so elusive and ephemeral: it runs through our hands, disappears into the earth, and goes wherever the wind takes it. Wilder can't keep the rain, but that does not prevent him from enjoying it. What a pleasure it is to share in this newcomer's pleasure.

But the relationships children form with nature are not always about pleasure; they are about survival. In recent years researchers in the fields of evolutionary and developmental psychology, among others, have made significant inquiry into the extent to which our childhood relationships with nature affect our physical, emotional, intellectual, and even our moral development. Children thrive in any number of environments, but there is a growing body of evidence that suggests healthy child development depends on direct—as opposed to vicarious— encounters with nature. In their introduction to *Children and Nature*, Peter H. Kahn, Jr. and Stephen R. Kellert make the point that "for much of human evolution, the natural world constituted one of the most important contexts children encountered during their critical years." Human children, like other animals, became what they are through interactions with the natural world.

Most of us would agree that from the time they are able, animals—perhaps most famously our beloved dogs and cats—have an innate drive to play. Their play takes many forms: stalking, chasing, and fighting, just to name a few. Young animals play as a way of developing the skills needed to survive the challenges they will face later in life, when they are on their own and must fend for themselves. The question I have been pondering is how and why children play in the ways they do, and what that play reveals about our past. That is, I wonder if play is adaptive among children. One of the ways we might attempt to answer this question is to ask whether children are predisposed to certain types of play.

When I look back over the last thirty-six years of my life and think about my relationship with nature and how that relationship has changed, I see that how I played was matched to my needs and to the stages of my development. I spent the first twelve years—well, almost twelve years—of my life in Caribou, Maine, a small farming community an hour-and-a-half south of the Canadian border. I was actually born in Idaho Falls, but I left not even a year later when my father accepted a job back east. When someone asks me where I'm from, I usually say I was born in Idaho and reared in Maine and Utah. In any case, the rural environment where I spent my childhood offered me the chance to play out my primal heritage.

Before we proceed, it may be a good idea to clarify my use of the word "play." Normally I would think twice before changing the meaning of a word, but when I open the dictionary and scroll down through the word's many meanings, I see that not one of them satisfactorily captures the scope of play as I think most children experience it. The current definition certainly doesn't describe all that I did as a child. Everyone who has lived through childhood knows that children do more, for instance, than play marbles and jump rope. Any honest adult knows that not all child's play is wholesome from the perspective of Western, "civilized" society.

As a society and as individuals we invest a great deal of time, money, and energy in encouraging our children to behave in certain ways (much of the time with good reason!), but based on my analysis of my own childhood, I wonder if we stand to gain more by exploring forms of play that our various interventions are designed to "correct." Considering that play is ultimately intended to familiarize children with the ways of survival, we should not be surprised if children get their hands dirty, or behave in ways not suitable for daytime television. Regardless of how long we've been on the planet, survival is serious business, and we have evolved to do what it takes to pass on our genes. Human children are deeply dependent on their parents for survival, but thanks to hundreds of thousands of years of natural selection, the moment they are born—actually, even before that time—they behave in ways that promote survival. Sometimes those behaviors land children in their rooms or on the evening news.

However, we would do well to temporarily suspend our preoccupations with "rightness" or "wrongness" of certain behaviors and instead focus on how certain kinds of play, insofar as they appear to promote survival, might be considered links to our ancestral past. Doing so promises increased understanding of why children behave in the ways they do, and if it happens that some of those behaviors are deemed undesirable, we may not only have a clearer sense of why—socially, individually, biologically—but also be better equipped to address them.

I realize this must sound like a disclaimer. But this concession is really just a comment on the complexity of my relationship with nature as it emerged and developed through play. I feel fortunate to have grown up in a place that enabled me to work out my relationship with the Pleistocene past. If the tendency to play is universal among children, then there must be some consistency in types of play. If children, owing to their shared origins, are faced with the same developmental challenges, we would expect them to exhibit particular forms of play to address those challenges. The play itself may appear different from culture to culture, but the skills attained through play will—if I am right—be generally uniform.

One of the advantages to growing up in a relatively natural environment is that play types are easier to identify. This is because in natural environments modern children encounter many of the same challenges faced by their Pleistocene counterparts. If, for example, children have an innate tendency to hunt, it is easier to observe if they live in proximity to prey species. Kim was quick to point out, however, that modern children also utilize the same play types in artificial environments. While earning her degrees in psychology and social work, Kim's primary focus was on early childhood development—she spent many hours interacting with children and watching them play. Although much of their play went on indoors, the children would often play hunting games. Sometimes they would hunt dinosaurs or an imaginary "bad guy," but they would just as often hunt each other, such as when they played hide-and-seek. Because learning to hunt would have been important to children throughout history, I imagine children from every culture engage in some form of hunting-oriented play.

I know hunting was certainly an important and difficult part of my life from the time I was about eight until I was fourteen or fifteen years old, or roughly around the time I became interested in girls. The movie character Napoleon Dynamite best sums up this transition from adolescence to young adulthood when he complains to his friend Pedro that he has no special talents, which is a problem, he says, because "girls want guys with skills: nun chucks skills, bow-hunting skills, computer-hacking skills." It was a funny line, but it underscores the intense drive boys (and girls) feel to craft themselves into desirable mates. In today's techno-saturated world, knowing computers is the Pleistocene equivalent to knowing how to make fire. Like Napoleon and other boys his age, I also became preoccupied with girls, but not before I had acquired my own skill set (excluding nun chucks—I couldn't wield them without hitting myself!).

Whether it was learning how to hunt, fish, or how to build shelters and fires, or learning the ways of human sociality, my interest in survival-directed play emerged in early childhood and, in some ways, continues to this day. It is important to keep in mind that the concerns and challenges of children are not the same as those faced by adults. Adult men generally do not play with matches or build forts out of chairs and blankets in their parents' living room. Having mastered fire and addressed their need for shelter, most adult men are busy pursuing mates or acquiring resources for their families, which they acquired, in part, because of their know-how. How we attempt to satisfy primal drives and preoccupations will change as we develop.

The decade or so when I hunted offers a useful example of this process. Of all the skills I acquired growing up, hunting was by far the most complex, because it involved interacting with the natural world and learning how to find and kill other animals. By the time I was thirteen or fourteen and was living in Utah, I had become averse to hunting. Actually, even now I still enjoy the hunt as a (mostly) catch-and-release fly fisher and as a birder—it is the indiscriminate killing I have long done without. It took several years and as many hunting experiences for me to develop a sufficiently robust, ecological morality, which hinges on the awareness that because all life is interrelated, we have a responsibility toward it. The act of killing was, therefore, oddly instructive.

When I recall my early memories of hunting, I realize how, even then, my urge to hunt and to kill troubled me. But now, a quarter century after that little boy stalked his yard with an air rifle and a pocket full of precious BBs, I think I might know why. The house on 318 Sweden Street in Caribou was built on roughly one acre and, much to our delight, featured a sprawling backyard bordered on three sides by trees and bushes. Whoever built the home had left a row of pine trees that split the yard in half. Scattered here and there were various trees and bushes, which made for excellent habitat for birds and hunting grounds for me. I am fairly sure I killed more than my share of birds during my time in Caribou, but for whatever reason—or for reasons I hope to reveal shortly—only two bird-hunting episodes come to mind.

Perhaps in an attempt to reconcile her own memories, my sister, Nicole, recently vacationed with her children on Long Lake in northern Maine. The lake isn't far from Caribou, so she drove down there for the day to look at the old house, which she found run-down, boarded up, and vacant. I haven't been back to Caribou since we moved to Utah nearly twenty-eight years ago. I guess I've never really had the desire. Yet when I heard about the condition of the house I felt a deep sadness. I guess I wanted the house to stay the way it was in my memory, so that if there came a day when I wanted to return, the house would be there. Or maybe what bothers me is how the wrecked and abandoned image of my childhood home serves as a metaphor for my final weeks in that house. Perhaps that's why I never felt my siblings' need to return. It seems that much of our desire to return to a place has a great deal to do with the circumstances under which we left it. Then again, people sometimes return to the scene of the crime.

If I had returned to Caribou, I would have traveled the old paths and visited the secret places. I would have learned the names of the plants and trees around our old house, as well as the names of the birds that inhabited them. I would have made amends, starting in the stand of trees on the east side of the yard, kitty-corner from the barn. Each year these trees bore olive-yellow fruits the size of cherries. Once the fruits—perhaps a species of plum—ripened, the trees filled with the chatter of gray-pink birds, their thick beaks perfect for breaching the tight-skinned fruit, which they ate down to the pit. At the time I did not know the name of the creature I was killing, and even today I cannot be sure, but given its habits, geographical region, and what I remember of its appearance, I think it was a species of grosbeak.

The fact that I killed these birds without knowing their name is revealing. I am reminded of that moment in *Silence of the Lambs* when—forgive the pun—Agent Starling listens to some audio evidence and comments on how the killer, Buffalo Bill, avoids using his latest victim's name. Doing so, she says, would force him to see his victim as a person. Presumably, the more we know about our victims, the more complex and difficult is the task of killing them. I don't imagine a lot of time is spent training soldiers to consider the humanity of the enemy (remember the Crow Indians). Contrary to the popular adage, then, it seems that when there is killing to do, knowing thy enemy simply isn't practical. Too many moral pitfalls, apparently—at least that is how I felt each time I watched a grosbeak shudder in the leafy sunlight and fall to the earth.

Inside the trees was a small clearing where I stood and waited for the birds to appear. I watched them poke and nibble at the fruit until a clear shot presented itself. Sometimes, after I shot them, the grosbeaks got caught in the branches as they fell. Most of the time I could dislodge them with a stick, but on one occasion the dead bird would not fall and it stayed up there until it was nothing but bones and feathers. Still, by season's end, I amassed a small necropolis of shallow graves, into which I placed the still-warm birds. Later, I returned and dug up the birds to check their bones. Once or twice, I returned too early and their bodies boiled with maggots and beetles. The smell of decomposition made me woozy. I felt woozy often in that New England town, where people made a habit of enriching their soil with fish heads, and animals of all kinds died quietly or violently in the dense vegetation, where, if it weren't for the stench, they would have remained unknown.

Our response to decomposition is interesting indeed. The smell repulses and repels us, but it also sparks our curiosity. We know something has died, but what and how, exactly? How might that knowledge be useful? I suspect a decomposing rabbit smells very similar to a decomposing deer or human. But a dead rabbit has different implications for us than does a dead deer or human. Is it possible that our need to know the difference might be adaptive? Whatever is capable of killing a deer would also be capable of killing us, so there might have been some incentive to determine the origin of the smell, if only to rule out the possibility of a large predator in the area. I guess we could have just fled the scene, but as we know from watching the popular TV show *CSI: Crime Scene Investigators* and its many clones, the scene of the crime offers invaluable information, including tracks indicating the direction of the predator's departure.

ii. Death's Rich Pageant

> The cemetery is an open space among the ruins,
> covered in winter with violets and daisies It might
> make one in love with death, to think that one
> should be buried in so sweet a place.
>
> —Percy Bysshe Shelley

My decision to bury the grosbeaks may have had something to do with the guilt I felt each time I shot one of them. Sounds plausible enough—but what, if anything, does guilt have to do with burial? Did I bury as a courtesy? As penance? For fear of being found out? None of these explanations hits upon my motive, so I wonder if there is another, more practical explanation, one that stems from the fact that humans have long been confronting corpses—human and otherwise. Despite varying beliefs about what, if anything, happens after death, I think most people would agree that burials are more for the living than the dead. Burial gives us closure, a chance to say goodbye, and an opportunity to "pay our respects." And it does not hurt our reputation if other humans observe us exhibiting these sensitivities.

However, these benefits don't account for why and how we came to bury the dead in the first place. While we normally associate "the dead" with humans, we bury other animals as well, and other animals bury, or cache, other animals (or their remains). We bury pets, albeit generally without the extravagance of human burials. Although we extend different courtesies to the corpses of our fellow humans than we do to other animals, our tendency to bury (or burn or bury at sea or launch into space or whatever) may ultimately be motivated by our evolutionary past.

Perhaps the practice of separating the dead from the living emerged as a way of dealing with the fact that decomposing bodies posed a danger to the living. Entomologists have observed this behavior in ants, which remove the bodies of deceased ants from the nest and deposit them some distance away in an ant "graveyard." Whether ant or human, carrion attracts animals. Granted, a comparatively small percentage of these animals would have been powerful enough to prey on humans, but other animals would have been attracted as well, including scavengers, vermin and other disease-carrying organisms. At some point in our recent history, it was no longer amenable to dispose of the corpses of our loved ones with so little fanfare, especially given that—for many religious traditions—the dead would need their bodies (or their spiritual equivalent) in the next life. Therefore, allowing the worms to dine on Grandma was no longer an option. Imagine if she had shown up in the afterlife in that condition! Thus, our various burial rituals enable us to slow or, in the case of cryogenics, thwart the ravages of decomposition.

In our effort to preserve the bodies of our loved ones (really, our memories of their bodies) without them becoming a risk to us, we select special sites and materials for their interment. Although it is certainly true that one's status in life and other cultural filters had and continue to have some impact on the type and quality of burial (not many of us would bury a grosbeak in a handcrafted box with satin lining), the tendency to separate the dead from the living appears universal. Available materials and cultural beliefs also affect where and how we bury. But whether we bury the dead in caskets, sarcophaguses, or pyramids, entomb them in stone, lay them atop platforms elevated above the earth, burn them, or lay them in a hole in the desert, each method is—among other things, of course—an expression of the same need to protect ourselves from the unseemly side of corpses.

Before humans had the means or the mythological motivation to construct containers for the dead, as groups of humans became bigger it would have made sense to bury the dead in one place so as to avoid creating multiple feeding sites for scavengers and predators. Presumably, by concentrating all the bodies in one area, those areas would have been dangerous places. Despite this apparent disadvantage, however, the decision to amass corpses in a single location, rather than to deposit them in various places, is really a no-brainer. Not only would it have been easier to describe and remember the location of the burial site, but concentrating the bodies would have significantly reduced the danger they posed.

Squirrels use the opposite strategy when caching food. In contrast to a grizzly, which is a powerful predator, the squirrel's position in the ecosystem doesn't afford much recourse if a larger creature decides to raid its cache. Rather than putting all its nuts in one place and standing to lose them all, the squirrel caches them in several places, ensuring he'll have a meal—that is, provided he doesn't forget where he cached his nuts. Obviously, humans don't bury other humans so that they may eat them later, but it would have been important for us to remember where we buried, and we certainly would have been faced with scavengers digging up the dead and creating an environmental hazard.

However helpful it may have been to construct burial grounds on the outskirts of the habitat (a practice that still continues to this day), there may have been at least some incentive to actively protect these sites. If early humans did in fact protect their dead from scavengers (many modern cemeteries are typically fortified with impressive fences, security guards, and electronic surveillance systems), it appears we have one more reason for amassing the dead: it's easier to defend one site than it is to defend several.

The current taboo against grave desecration, which generally entails digging up and disgracing the dead, might reflect this ancestral behavior. But usually desecration involves more than exhuming the dead. Humans are horrified by the idea of being ingested by another animal. I think this is what made grizzly bear enthusiast Timothy Treadwell's death so frightening: when the float plane pilot first spotted Treadwell and his companion, a grizzly was hurriedly feeding on Treadwell's body. It's one thing to be killed, but to be killed and then eaten! On my personal list of terrible ways to die, being killed and eaten by a grizzly ranks at the top, right under being killed and eaten by a shark. Our aversion to becoming another animal's meal is what makes desecration such a cause for unhappiness in the living. We simply don't respond well when some low-down, graveyard dog and a legion of fellow scavengers ravage the bodies of our loved ones.

Incidentally, this familiar description ("lowdown") of the dog's orientation to the earth belies the related notion that animals whose bellies are close to the ground are that much further away from God. Humans are in good standing because, well, they can stand. Their heads are in the clouds and their bellies ride high above the creatures of the earth and the earth itself, which is what, finally, God-aspiring humans must transcend. I guess some folks have never seen a giraffe! And let us hope for the snakes of the world! But I digress.

The cemeteries themselves may offer the best insight into primal heritage. Cemeteries differ from place to place, but generally they are among the most controlled human-made environments. And, as I think Shelley implies in the epigraph at the start of this discussion, cemeteries inspire conflicting emotions. Our mixed response stems from the fact that burial places were dangerous yet "sweet" places. Because burial places may have been dangerous to early humans, perhaps natural selection would have favored humans who offset these threats by selecting safe environments in which to bury—we would have sought and created the characteristics of a safe environment. Over time, these characteristics would have become the basis of why we respond favorably to some habitats and not to others. Modern cemeteries should, then, be constructed in ways that reflect these requirements and maximize our feelings of safety and well-being.

Cemeteries are fairly rudimentary in their construction, but they are appealing—perhaps because their design features are saturated with primal significance. Perhaps the single most defining feature of cemeteries (other than daytime visiting hours) is their openness. When traveling in a dangerous area, most of us would prefer less mystery, not more. In regions where water is available, this openness is often complemented by short, sprawling, green turf. Both conditions—wide-open views and short grass—appeal to us for several possible reasons. One possibility is that they would have aided our ancestors in the detection of predators, but I think there's more going on here. Research has shown we prefer environments that resemble our original habitat, i.e., the savannas of Africa.

I wonder if, in addition to its length, the deep green color of the grass is also an important part of our cemetery psychology.

Although the color of the grass may not have anything directly to do with ensuring our safety, it does seem related to the deep aesthetic appeal cemeteries have for us. Gordon Orians has proposed one possibility for why we might use flowers in various rituals, including weddings and funerals. He suggests that flowers may have a special importance to us because the blossoms of certain flowers signal fruiting trees and plants. How exactly that importance translated into adorning graves with flowers and giving flowers as gifts remains unclear, but I think Orians' theory might help us understand our attraction to deep green grass, its color and, possibly, even its smell, which for many of us is as pleasing as a fragrant flower.

The savannas of Africa are, then, an ideal place to begin looking for ideas as to why grass appeals to me. With the savannas firmly in mind, I went into the backyard and gazed at the new sod I laid a couple of weeks ago. It is with some embarrassment that I admit to having laid sod here in the desert of all places, but I console myself by noting that Wilder needed a safe place to crawl and play. As long as I am being honest, I must also admit to feeling a pleasure that goes well beyond having provided a safe environment for Wilder. Much to Kim's amusement, I must have commented a dozen times on the grass's beauty. She often caught me looking out the windows and admiring it. "I guess I should be thankful," she said once. "You could be ogling the neighbors."

Given my attachment, I was nervous the first time I mowed the lawn. I had to be careful not to cut it too short or the grass would die. But as the blades of my electric mower began to hum, slicing the delicate leaves of grass with the neatness of barber scissors, my worries ended and off I went, down the timeline of grass. First I was transported to Utah and the house on 9039 Huckleberry Court, where I saw sod and sprinkler systems for the first time. I thought back to Sweden Street, and to grass so long it had doubled over like a kid who just got punched in the stomach. Finally, I alighted in my grandparents' yard in Idaho and revisited my earliest memory of grass—how it concealed the broken beer bottle I knelt on while playing Cowboys and Indians. Grass and I go way back. In fact, I can't remember ever having lived without it, even here in the arid Southwest.

The Sonoran Desert is a long way from the savannas of Africa, but neither part of the world is known for its rainfall. When the rains arrive, it is a welcome and crucial event. When all we need to do is turn on a faucet to get water, it's easy to lose sight of rain's importance to the ecosystems that depend on it. This dependence would have been especially apparent on the savannas of Africa, where the rains truly meant the difference between life and death. Maybe one of the reasons we have such a powerful response to the color green in general and to green grass in particular is because the color signaled that the rains had come. We would have begun to develop positive associations with the color green. A lot of grass means a lot of water, and a lot of water means a lot of animals, which in turn means a lot of potential food. However we organize this web of interdependence, as an environmental cue, green would have meant (and still does mean) "life is good!" It should come as no surprise that the environmental movement has come to be known as the *green* movement. In any case, this might help to explain why sod covers tens-of-thousands of acres in the Sonoran Desert and elsewhere in the arid Southwest. It could be that we are unwittingly attempting to recreate characteristics of our first home.

On the savannas of Africa, most of the larger animals would have been herbivores, herds of them, grazing on the lush grass. If part of why we experience pleasure when looking at green grass is because we associated the grass with prey, then any information related to the acquisition of that prey would, presumably, also inform our sense of pleasure. Although I have not had the opportunity to smell the dung of wild gazelle, wildebeest, zebra, giraffe, elephant, or the other herbivores that wandered the savannas, I don't find the sweet, grassy smell of horse droppings altogether disagreeable. Must be my German ancestry! Like the droppings of savanna herbivores, horse droppings are basically condensed balls of cut grass.

I suppose it's possible that hundreds of animals simultaneously grazing could release the characteristic sweet odor of cut grass. More importantly, perhaps, is that the droppings would have been valuable pieces of information, not only in signaling the presence of prey, but also its nearness, which would have been judged by the dropping's moistness and pungency. Wondering if I were on to something, I asked Kim if she liked the smell of horse droppings. I should note here that Kim 1) knows me better than anyone on the planet and 2) is well-versed in Darwinian evolution. Otherwise, I might be writing this book in a padded cell. She immediately responded by saying "No, not especially." I was about to say goodbye to this line of thinking when I realized there may be a reasonable explanation for why Kim didn't share my olfactory enthusiasm for horse dung. For most of human history we were hunters and gathers. As the physically more powerful sex, males did the majority of the big game hunting while women gathered and hunted smaller animals. Perhaps females don't have the same reaction to the smell of herbivore droppings because generally they weren't the ones hunting the animals that produced them.

This hypothesis reminds me of an interesting study done by Sarah Tishkoff and Brian Verrelli. Using 236 adult male DNA samples originating from Africa, Europe, and Asia, the two studied variations in what's known as the red opsin gene, which basically accounts for how humans perceive the color red. Tishkoff and Verrelli's findings show that women have an enhanced perception of the color red, which (they speculate) would have been important given their role as gatherers, an activity that depends on the ability to discriminate colors, especially those of fruits, insects, and other edibles. One of the things that makes this study so compelling is the size of the data set: acquiring 236 DNA samples from different human populations and parts of the world is no small feat! Out of curiosity and, I admit, to determine if my speculations were at all supported by other peoples' experience, I asked twenty friends (ten men and ten women) the following question: "Do you respond favorably or unfavorably to the smell of horse droppings?" Before I am accused of having too much time on my hands, I should point out that this small sampling yielded some interesting results.

Given Kim's and my differing responses, I predicted that, on average, women would respond unfavorably while men would respond favorably to the smell of horse droppings. The men were evenly split. Initially I was a bit disappointed by the men's percentage, but then I tallied the women's responses. All ten women responded unfavorably! Compared to the women's 100% agreement, 50% agreement among men now didn't seem so insignificant. As I thought about it more, I was surprised that any of the men admitted to responding favorably to the smell of horseshit. Imagine if I had asked the same question, but instead of "horse droppings" I said "just-cut grass." In terms of what's socially acceptable, for a man to admit to liking the smell of cut grass is as harmless as admitting to liking the smell of cotton candy. But for a man to admit to liking the smell of shit—even horseshit—well that's just crazy talk!

I don't wish to suggest that all the men who answered "unfavorable" were being dishonest, just that the socio-scatological implications of the question might have undermined accuracy. I sensed that some of the men might be uncomfortable with the question. (Next time I'll be sure to craft a question that doesn't raise any eyebrows.) I therefore advised them to avoid analyzing the question once I had asked it. Interestingly, the men who thought the longest were twice as likely to respond unfavorably as those who answered immediately or with little thought. Certainly, this poll leaves much to be desired, and I'm still not sure why women respond one way to the smell while men respond another. It might just be because women generally have a more sensitive sense of smell than men. Still, the numbers are interesting. Perhaps one day some ambitious, well-funded evolutionary psychologist will conduct a comprehensive study on the olfactory response to herbivore dung. Keep the dream alive!

If this exploration of why and how we might have come to bury our dead in the ways and places that we do seems a bit far-fetched, consider how these phenomena most often appear in popular culture. Cemeteries have long been a staple of the horror film genre, particularly fallen cemeteries, or those that are overrun with vegetation. The moaning and maggot-laden dead emerge from these sites with a hunger for warm human flesh. Why burial places are so frightening, and figure so strongly in the human imagination, must seem fairly obvious from the perspective of popular culture. But movies aren't the only medium that features cemeteries as locations of terror. Just the other day I was channel surfing, ever on the lookout for data, and I happened to catch the trailer for a new TV show called *Nightstalker*. The trailer of this supernatural series features the following voice-over: "What if everything we fear is real?"

The question is comically rhetorical for anyone who is biologically literate. For whether they are rational or not, all fears are real to the extent that they have a biological origin. Cinematic depictions of what we fear—including cemeteries and the dead that fill them—frighten us because, even as imaginative elaborations, they are still rooted in the real danger these phenomena posed over the course of human history. In fact, horror films that do not in some way tap into this primal reservoir fail to frighten us.

Based on my experience with and study of the horror genre, horror films share certain characteristics, particularly with respect to location and monster-type. In addition to cemeteries, deep woods and open water are two of the most familiar locations (*Blair Witch Project* and *Jaws* are obvious examples), but so too are dark, unfamiliar houses. However much we might be inclined to think otherwise, it's a short step from a cultural analysis of these films to an evolutionary one, which proposes that these locations might frighten us because we have always associated them (or elements of them, e.g., the unfamiliar) with danger.

The motion picture industry has also seen its share of monsters over the years, but again we can point to certain types, including fantastical creatures, extraterrestrials, nonhuman animals, and humans. I would argue that the degree to which each of these types frightens us depends on how well they evoke Pleistocene hazards. Because humans were and continue to be the single greatest threat to our mortality, humans or humanoids should constitute the greatest percentage of monsters. (The next highest percentage goes to nonhuman animals.) A glance at the many Top 50 horror movie websites supports this conclusion. Horror films that attempt to cast modern technology, for instance, as the monster simply lack the potency and appeal of those terrors that have confronted us since the emergence of life. The notion that our emotional responses to these locations and monsters are, like the locations and monsters themselves, merely cultural artifacts seems unlikely and, quite frankly, boring.

iii. Morality, Grief, and Waste

Compared to the pageantry, solemnity, symbolic significance, and mystical undertones of many modern day burials, I know that these hypotheses must appear to threaten the sanctity of everything we hold dear. Perhaps we feel they diminish our sense of grandeur, as if exploring the origins of our behavior meant we were any less extraordinary, or less committed to our dead. Our rituals are so luminous and symbolically weighty that wondering how they came to be must seem like a sacrilege. One of my own favorite examples of this funerary richness aired on PBS a few months ago. The program featured the discovery of naturally mummified children found on a mountaintop. The children were members of a horse culture, the wealthiest of whom—a young girl—was entombed in stone with nine horses, each of which had been carefully bridled, saddled, and bludgeoned! Given such grandeur and spectacle, we may fear we will be diminished if we start asking questions.

Obviously, very different sensibilities inform our various expressions of burial and why they occur, including the evolutionary hypothesis. (I imagine one of my slain grosbeaks traveling down Sweden Street, wings folded across her chest, her prostrate form rocking slightly with each pothole, but otherwise riding motionless within a case of glass.) Even so, these understandings—the proximate and the ultimate—are not incompatible and therefore need not be viewed as inimical. It's difficult to suppose that we might find the same underlying reasons for burying a nameless bird, a pet, and a beloved human. And understandably so—for now we must extend to other creatures the moral awareness typically reserved for humans, and that's when things start to get complicated.

This may be a good place to repeat the E.O. Wilson quotation I included at the beginning of this book: "Humanity is exalted not because we are so far above other living creatures, but because knowing them well elevates the very concept of life." Aren't we, in effect, elevating the concept of life when we acknowledge that our own behaviors, including our most sacred rituals, might stem from a time in history when our needs and the needs of other animals were essentially the same? I think this is why I've always remembered killing those birds as they went about the daily business of feeding themselves. Although my shame has diminished over the last quarter-century, my memory is vivid right down to the errant feathers and the dead leaves, and to how easily the cool earth gave way to my fingers as I dug each tiny grave.

After a decade of hunting and killing, I finally reached a level of moral awareness and maturity that prevented me from killing needlessly. But even before that point I knew that killing those grosbeaks was wasteful, unnecessary and, therefore, wrong. Each time I killed for the sake of killing, I broke the ecological contract and killed myself in small ways. Humans would have benefited by developing a measure of restraint when it came to hunting and killing. Our early ancestors were, then, the first ecologists by default. Apart from the likelihood that the majority of their hunts were unsuccessful, their lifestyle would have required deep knowledge of the animals they depended on in order to survive. If there were any merit to these claims, it would mean that ultimately nature provides us with a sense of biological morality, which develops in relation to our environment.

When we look at hunting cultures we find that the adults generally guide children as they acquire skills and negotiate the inevitable moral implications of killing another animal. Over the years I have known a few hunting families, and in each of them the tradition of hunting was something the father shared with his son and, sometimes, with his daughter. Instruction mitigated the impulse to kill. Through this process, most of these boys became sensible killers, which shows that even a modest level of involvement on the part of older humans—father or grandfather, for instance—clearly has an impact on the development of this nascent moral sense.

My last weeks on Sweden Street, during which I killed the grosbeaks for the final time, were difficult for several reasons. By that time my mother had already been in Utah for a month or so, where she hastily prepared for my brother Christian and me to join her and my sister Nicole. That left my father, brother, and me to fend for ourselves. Christian developed an interest in guns after we moved to Utah, but during our time in Maine I don't think he ever laid hands on one. Although he is only a year older than me, that year made all the difference in terms of our interests. He must have been about thirteen years old the year we left Maine. Prior to that time we were fairly like-minded, but once he became a teenager, it was as though a switch had been flipped and his attention turned to girls and the various means and ploys used to attract them.

Not that my father was a hunter, but I don't remember him being around much as the time of our departure drew near. In fact, I do not remember the three of us being together until we were standing at the airport gate and my father scooped my brother and me into his arms, kissed us, and then disappeared into the crowd. I don't think any of us realized the depth of that goodbye. My brother, father, and I were not much different in the sense that we did what we could to prepare for the dissolution of our family and life as we had known it. But rather than preparing together, we prepared apart. My father was the only one with the ability to recognize this problematic and familiar expression of grief. But who can blame him for wanting to absent himself from the catastrophic (social and biological) failure of his life at that moment?

The grief of losing one's children is a powerful emotion, so the incentive to disconnect from such a destructive feeling is strong—perhaps because in the ancestral environment it was crucial to get on with life, have more childrenn and transmit more genes. But of course losing children to death and losing them to distance are two different things. We live in a time of global communication and transit. In the context of the modern world, one could make the argument that my father (and my mother, in her own way) did us all a disservice by permitting his genetic predispositions to run roughshod over our familial investment and needs. Perhaps one day *foresight*, rather than *hindsight*, will be 20/20.

This disconnect culminated when he remarried and had another child. If there were any chance that we would develop a meaningful, albeit long-distance relationship, it was undermined by the creation of this new family. This phenomenon is referred to as parent- offspring conflict, which John Alcock describes as "the clash of interests that occurs when parents can gain fitness by withholding parental care or resources from some offspring in order to invest in another round of reproduction at a time when existing offspring would benefit from receipt of the investment." I wonder how things might have been different had my father, a psychologist, been more aware of evolutionary theory. Perhaps he could have used that knowledge to help him more effectively respond to (or even prevent) the crisis of his life. Instead he became an evolutionary psychologist's prediction. He became a statistic.

In any event, my father, brother, and I each dealt with—or avoided—the grief of upheaval in our own way. Without the guidance of someone who could help me understand the implications of killing, I was free to act out my primal interests unchecked. For the time being, then, my nascent moral awareness slept and would only gradually awaken until I reached a level of development that enabled me to favor a more ecologically informed approach to other animals, which to this day continues to develop. I suppose this is the difference between "mindful" and "mindless" and "needful" and "needless" killing. As an eleven-year-old child, I did not have the cognitive ability to articulate this crucial distinction, but I recognized it when I saw it.

The human aversion to waste is ancient and unmistakable. I think this is what I felt each time I killed a grosbeak. Apart from the information gleaned from examining their tiny corpses, I knew that killing those birds was without purpose. It would have been one thing had I eaten them or used their feathers or put them to some practical use. Instead I studied them until they were cold and then placed them in the ground. What has troubled me the most over the years is that I kept on killing even though I intuitively knew that doing so was fundamentally untenable. Had someone been there to guide me, things might have been otherwise: I may have found a more sensible way to express and manage my instinct to hunt. But is it possible that my willingness to kill the birds, and the degree to which doing so registered on my innate moral compass, might also have something to do with our ecological relationship as it has formed over time? Perhaps if we revered the animals upon which we most depended for our survival, an animal's nutritive and material value would have led to its moral and symbolic value.

Early spring in the woods of northern Maine is a dangerous time for the snowshoe hare. The hares retain much of their white winter coats even though much of the snow has melted, making them vulnerable to predators. A couple years before we left Caribou, my mother's boyfriend Rand drove my brother and me out into the woods to hunt snowshoe rabbits. We came to a house deep in the woods that was so full of holes it looked like 10,000 woodpeckers had descended on it. I think Rand let me shoot at it with his .22 rifle. By the time the sun was going down and slush on the forest floor was turning to ice, Rand was the only one to have shot a snowshoe. He held it by the rear legs and slung it over his shoulder. He wore a yellow rain slicker, and for a while the rabbit's blood ran down it. Walking a few steps behind him, I could see where it dripped into the snow.

It was full dark by the time we got home. My mother met

us at the door and I gleefully held up the rabbit in the porch light. Rand skinned the rabbit that night after I had gone to bed. In the morning I went out on the front porch and found the pink and densely muscled hare soaking in a large pot of water. The porch boards were cold on my feet. I must have repeated this ritual for about a week until someone finally threw out the rabbit. I can still remember the disappointment I felt. I knew something precious had been wasted. Had I not seen the snowshoe alive in the woods, nor known that it was meant to be eaten, I may not have had quite the reaction I did. But I had seen it, and so its life made its waste all the more offensive. To some extent, I had also witnessed the grosbeaks living their lives high in the crowns of those trees, but the snowshoe was an unmistakable opportunity for food. Back in the day, our ancestors did not know when they would have another chance to eat. Catching an animal would have taken significant energy. In the context of these variables, then, wasting was quite literally unthinkable.

This aversion to waste would have extended to other parts of the animal as well, especially bones and fur. Although I became somewhat of a bone collector much later in life (the skull of a young peccary adorns my yard, and two others sit atop my office bookshelf), I don't remember ever encountering any bones. However, I do recall briefly having in my possession a mink fur and a deerskin. I'm not sure how I came to possess them, but the mink fur was silky and had the smell of treatment, whereas the deerskin, though soft and supple, had yet to be treated. I tried to sell the mink skin for ten dollars by placing it with a sign in one of the two oak trees that shaded our front yard. I remember my puzzlement when the days came and went and no one stopped to buy it. After all, this skin was valuable, wasn't it? I figured it could be used to line moccasins, or be fashioned into gloves. After three or four days without interest, I took down my sign and brought in the mink fur, where it inevitably disappeared into the sinkhole of time.

The deerskin met a similar fate. Initially I laid it on my bedroom floor, but it had a tangy, sour smell, and so after a little prodding from my mother I put it in the barn. The skin stayed in the barn throughout the fall and winter, until it froze and was ruined. I would not describe any of the experiences I have recounted here as especially painful or pleasant. And yet I distinctly remember them.

Adaptively speaking, we remember painful and pleasant experiences so that we will either avoid or attempt to reproduce them. But the fact that neither of these reactions describes my response suggests another dimension to the human memory, what might be described as an eco-moral dimension that encompasses pleasure and pain but is characterized by a profound aversion to waste. The eco-moral dimension of memory stems from humanity's recognition that we are truly and deeply dependent on other animals and on the environments in which they live.

iv. Fort-Building

Children who were not motivated to hunt and kill in the ancestral environment would have been severely disadvantaged and, ultimately, selected against—whereas children who expressed an interest in these activities would tend to live to reproductive age and thereby pass on their genes. Therefore, whether modern children hunt (and sometimes kill) dinosaurs, imaginary "bad guys," grosbeaks, or each other, they have their Pleistocene counterparts to thank for their interest in hunting-oriented play. Knowing how to hunt and kill was crucial to survival, but knowing how to find and construct shelter would have been just as important.

I think this is one of the reasons why, when out in nature, we are curious about anything that even resembles a cave, and why we are generally quite skilled in detecting them even though most of us haven't had to inhabit a cave for thousands and thousands of years. Always on the lookout for suitable accommodations, we seem to have our own built-in home finder. However, because there are only so many caves that were well situated and safe enough to inhabit, a good cave would have been hard to find, that is, assuming there were any caves to be found at all. Prodded by the threat of the elements and of predators, hominids that were resourceful enough to construct their own shelters would have surely outlived and outreproduced their more idle neighbors. As the descendants of those early humans, modern children often express an interest in shelter construction.

In their recent essay "The Ecological World of Children," Orians and Heerwagen cite studies that show a preferred play-activity among children is the construction of shelters and refuge. Although I have only touched on what is already a substantial and growing body of research, my speculations are supported by scientific data, and not just by my experience with practically every child I have ever encountered. That my relationship to hunting and killing would change over time would not surprise Orians and Heerwagen. Their own work charts a similar progression in children's interest in finding refuge and constructing forts or shelters. This progression corresponds to the child's stage of development, in which case very young children, who are still dependent on parental care, would tend to construct shelters in close proximity to the parent, whereas older children, who in the ancestral environment would have begun to hunt and were in a much better position to fend for themselves, play and build shelters farther from home.

It's not only with respect to proximity that Orians and Heerwagen's work describes my experience with fort-building: their finding that children prefer shelters with certain characteristics does, too. Regardless of their age, children construct or inhabit shelters that offer them privacy with the ability to see out. For young children, it is important to maintain sight of the parent, whereas older children, who typically construct shelters on the periphery of the home site, would require visibility for the detection of predators or enemies. Geography has a lot to do with the types of forts or shelters children construct, but generally, they each reflect our preference for refuges that enable us to see without being seen.

I remember well my own history with finding and creating safe places to play—first in my immediate and later in my surrounding environment. The first time I remember finding refuge for myself was at my grandparents' house in Idaho Falls. I must have been about four years old at the time. I don't recall being especially fond of my grandparents' backyard. I'm sure there were several reasons for this, including what I recall as the somewhat strained relationship I had with my mother's parents, especially Iris, who would often tend us while my mother was out. More than that, however, was the fact that the yard wasn't mine.

I could have handled the unfamiliarity factor if it weren't coupled with the breach in the back wall. The stone had been cut away so that the garbage cans could be placed in the alley. When the cans weren't out, there was nothing to prevent intruders from entering the yard. I couldn't have spent more than a few weeks of my life in my grandparents' backyard, but I remember that long alley and how it was overrun with weeds and shadows. When my sister, Nicole, was about seven years old she was playing near the breach and a man approached her and told her he'd give her a nickel if she pulled down her panties. I asked my sister about the incident just the other day. "That nickel," she said, "it's up to a dollar now." We both laughed, relieved that nothing more had happened—my mother saw what was going on and rushed outside. The man ran off. Lucky for him. Years later my mother would say, "Do what you want to me, but mess with my kids and I'll kill you."

After she divorced my father, faced with the challenge of supporting three kids, my mother did what she had to in order to survive. My mother was and is a smart and beautiful woman, so she had no trouble attracting males, despite the fact that she already had three kids in tow. In fact, by the time my mother had had her fill of men (she's been solo for the last eighteen years), she had been married five times and had two or three boyfriends in between. Considering that only one of these males was my biological father, and that my siblings and I spent several years of our childhood in the presence of unrelated males, life was pretty good. But things could have easily been otherwise.

My mother was raised in the Mormon Church, and even though she has not been an active member for decades, she's still a little embarrassed by her conquests. Of course, she wasn't taught to call them "conquests." That would not have been the right brand of empowerment. Instead, what little embarrassment she still feels stems from the doctrine of sin. But I don't look at it that way. I know better. She did what she had to and still managed to keep us safe. Had she been any less of a woman and mother, and not made it perfectly clear to her suitors that the way to her was through her children, who knows how some of those unrelated males might have treated my siblings and me.

Given their responsibility to their offspring, women have always had to be especially resourceful when it comes to safety. Like every other resource on Earth, men are (individually, collectively) finite. These days it is not uncommon for men to live for seventy or eighty years—which in many cases is longer than what is in their own, their family's, and the planet's best interest. The men who lived even as recently as 1,000 years BP, not to mention 100,000 years ago, lived until they were twenty-five or thirty, if life went their way. With my father out of the picture, my mother did what hominids in her position have been doing for thousands of years: she found another male. Apparently the risk posed by unrelated males was less than if she had tried to support us on her own. Eventually she would get to the point where she no longer needed the help of men, but in the meantime she was ever vigilant of our safety, and so took her place among the millions of animals who have borne offspring.

I did not always know that my mother was nearby, and surely there were times when she wasn't. Perhaps this helps to explain why I searched my grandparents' yard for places to hide and play. Their house had been built at the turn of the nineteenth century and featured several large trees, shrubs, and hedges that had been growing for nearly as long. At that age, I hadn't developed the musculature necessary for tree climbing, so instead I would burrow tunnels throughout the hedges that lined the back walls. I preferred the large hedge in the northwest corner, the bulk of which grew in the corner and then gradually tapered off on either side. Because of its size, I was able to make several tunnels leading in and out of the hedge. True to Orians and Heerwagen's prediction, what made this shelter appealing was that I could see out but no one could see in, due to the low profile of the tunnels. Any animal larger than a five-year-old little boy on his hands and knees would not easily breach the inner sanctum of my hedge.

Unfortunately, as anyone knows who has ever lured a creature from its hiding place, a fortification is only as impregnable as the will of its occupant. Like other children I knew at the time, including my brother Christian, I was fascinated with fire. I remember the excitement I felt whenever I found a book of matches in my grandparents' house. The more matches there were, the more excitement I felt. I would stuff the book into my sock and steal away to the deepest part of the hedge, where I would place the match on the strike pad and carefully fold the book cover over it. I remember liking the "snap" of the igniting match, and the sweet, raw smell of sulfur that stung my nostrils before finally terminating in flame. Once I had sheltered the flame from the breeze and any wayward eyes, I conveyed it to small piles of dead leaves that burned into tight curls of ash.

During my early teens I met boys who were recklessly obsessed with fire, but I have always had a pronounced respect for it, have always prided myself on making a fine fire and doing so according to the "one match" rule. I would like to think I naturally developed this respect, just as children have always done. But, alas, my uncle might be partly to thank. I don't recall the exact circumstances, but I do know he caught me and Christian playing with matches a few times. The last time he caught me I was under the hedge—except I didn't know he had caught me. He was a big man back then and I remember seeing his big sneakers in the grass and his light blue corduroys up to his shins. I quickly extinguished the small pile of leaves and wafted away the smoke.

"Come on out and I'll take you and your brother for ice cream," he said, rather sweetly. I liked ice cream even more than I liked matches. When I think about his words now, however, I realize something wasn't right from the very beginning. Normally my uncle spoke to me in a rather disdainful, booming voice, like he was talking to some forlorn child of the damned—Christian and I had not been baptized into the Mormon Church. We were the sons of an atheist father and of a mother who, the more she looked at how her own family embodied the church's teachings, did not like what she saw.

When I add our damned status to the well-established fact that my uncle was just a flat-out mean son-of-a-bitch, crazy as a shithouse rat, I realize both the extent of my trust and the power of my sugar tooth. I left the matches under the hedge and out I crawled into the cab of his old Ford pickup. My uncle still had to convince Christian to join us. Ice cream or not, Christian was skeptical. Imagine that: a six-year-old skeptic. We had learned early not to trust this man, this stranger in every way but name. Christian was less willing to let bygones be bygones. After a few minutes of intense negotiations (perhaps my uncle sweetened the deal with the promise of a double-scoop and candy sprinkles) we were on our way.

I have replayed this scene many times: a promise is made; a destination is decided; travel ensues; a fork appears in the road; the way is east, but the man heads west; the old truck heads out of town; the two children are afraid to ask where they are going. One of them does; the man breathes through his nose and does not speak; the truck jumbles down a dirt road; the boys turn and look back; a piece of garbage swirls and falls, swirls and falls in the bed of the truck; the smallest child can see where they left the road; the trail of dust is already gone; the boys look at each other, as if to say *no one knows we are here*. The man parks in the shade of an old tree and turns off the engine. "This is what you get when you play with matches," he says, and proceeds to burn our fingers. I remember my uncle's massive hand holding mine and applying the match to the tips of my fingers. My brother and I didn't tell my mother about it until years later. I told her I wanted to see him suffer. "You do, do you?" she asked as she knelt in her garden. "That's easy. Just look at him."

Compared to the real and perceived threats posed by my grandparents' place, my siblings and I had very little to worry about back home on Sweden Street. That was our turf, after all, and my parents were still together. Actually, they weren't *together* together, but they lived under the same roof: my father had a room upstairs and my mother had a room downstairs. However unusual the arrangement, up until about a year before we left Maine, my siblings and I enjoyed the most security we would ever know. But when it was clear to everyone that my mother intended to leave and take us with her, things took a turn for the worse. We all suffered from the fallout, but my older sister Nicole suffered the most. Although I've always thought of Nicole as my sister, she is really my half-sister. My mother was pregnant with her when she met my father, and about a year-and-a-half later my brother Christian was born, and a little over a year later than that, I followed. To this day, Nicole has not met her biological father. My father is the only father Nicole has ever known.

This type of reconstitution reminds me of a passage from Carl Sagan and Ann Druyan's book *Shadows of Forgotten Ancestors*. The narrative comes from the chapter called "Gangland:"

> *The Big Guy, he's nice to me. One time my kid's watching us in the act and tries to stop us. He climb on, hitting Big Guy with his little fists. Big Guy, he don't touch him. He think it's funny. He don't hurt my kid. He don't hurt me.*

Given the fact that parents, particularly males, generally invest more in children who are biologically their own, one could argue that my father behaved admirably when he courted and married my mother. It takes determination to resist the selfish tendencies of the Pleistocene mind. If my mother had any doubt about my father's commitment to her and to her unborn child, it must have disintegrated the moment my father reassured her that she should not and could not raise my sister alone. What a primal relief! Big Guy may have spared the "kid" because he thought it was his own. But my father was under no such delusion. Still, for a time he was like one of the many parents—male and female—who throughout the history of the species have managed to overcome their biological bents in the interest of love and family.

Whether or not my parents ever actually "loved" each other seems iffy at best, which is to say they never even came close to developing that kind of relationship. When I asked my mother why she left my father, she said she never felt like he loved her. But it takes two to tango: my mother told me that she didn't know why, but shortly after their marriage, she was no longer attracted to my father. However tempting it might be to conclude that her lack of interest was inspired solely by my father, hormonal changes in my mother's body could have played a role as well. My parents were married in March, not two months after Nicole was born. Not the best time to inaugurate a sex life.

Apart from the soreness that usually attends birth, immediately following birth the costly lactation process begins. Human females are similar to other mammals insofar as they are not as sexually receptive when they are producing milk. This is why, for instance, after annexing new territory male lions will often kill the offspring of ousted males. Given the generally long-term dependence of mammalian offspring on their mothers, this hormone-induced stalling of female sexual receptivity evolved to ensure that existing offspring would be sufficiently grown before more offspring are produced.

This arrangement worked out well for Nicole and my mother, but it also meant that my father was going to have to postpone "helping" my mother. My father would later admit that really he was attempting to help himself: several years after their divorce he flat out told my mother he married her because he wanted to have beautiful children—she was "good stock." Perhaps, then, his initial interest had nothing to do with being a "good" man. Maybe he was just being lead around by his genes, which did what all genes do if conditions are right: survive at all costs. Nothing unusual about that.

Trouble was, Nicole did not bear my father's genes. Once he knew that my mother was leaving him, his mettle was tested. In the end, my father could not overcome the grief of being separated from his children and the selfishness of his genes, and so in the process of ending life with my mother, he abandoned my sister, even though he had been the only father she had ever known. No wonder Nicole left Maine months before my mother. It was only in retrospect that I realized this aspect of my father and sister's relationship. And I suppose it doesn't matter. It doesn't change things. Certainly not for my sister, who I doubt will ever revisit that time in her life. As Hurstwood in Theodore Dreiser's *Sister Carrie* would say, "What's the use?"

In some ways I can relate to how she felt, and I'm sure my brother can, too. Although my mother had our best interest in mind when she left for Utah, I still struggled with feelings of abandonment. I'm convinced that the only reason I was able to work through them was because my mother assured me with letters and the occasional phone call that we would soon be together. I don't know what, if anything, my father said to Nicole the first time he took my brother and me somewhere without taking her. I'm thinking he didn't say much of anything. Without recourse to reason, the only thing he could do was remain silent. I'm not saying he got off easy—he didn't. I'm saying that in the wake of that first silence, whatever it is in us that makes us trust began to die in my sister.

Thus, in addition to being places where my siblings and I could shelter ourselves from imaginary foes and monsters, the refuges we constructed in and around the house on Sweden Street helped us cope with the very real emotional toil associated with the dissolution of our family, while preparing us for a life on our own. The notion that we experienced this tension more acutely inside the home is perhaps reflected in the wonderfully elaborate forts we would construct out of chairs and blankets in the living room. Very often the forts were so large that they filled the room, and like the snow forts we would construct each winter, and the ruins of the ancient ones, they were comprised of several "rooms," each with a specific function, including food storage, cooking, and sleeping quarters.

Using an evolutionary framework, Orians and Heerwagen predicted and found support for the tendency older children have to seek shelter farther and farther away from the home. This tendency is correlated with the child's development, particularly with respect to increased physical strength and mobility. Thus, the older and more able they become, the more likely they are to strike out on their own, going farther and farther, until finally they are skilled, knowledgeable, and mature enough to set up shop, attract a mate, reproduce, and hopefully live a long and happy life. A month or two before we left Maine I was already showing an interest in wandering. During my last week in Maine I would wander farther than I ever had before. But before that time I was content to search for and construct shelters in our barn.

Although the barn was attached to the house, and so remained within the purview of my parents' protection, we had no trouble finding privacy. As far as barns go, ours was very elaborate. It had three levels, each with several stalls. I can't say for sure how many stalls there were, but I'm thinking somewhere on the order of twelve, including two on the lowest level. Were I unfamiliar with the biological underpinnings of our dwelling preferences, I might be surprised that we rarely spent time playing in the stalls, which, because they were self-contained, would seem to have been ideal forts. But therein lies the rub. For unless one were strong and foolish enough to attempt ascending or descending the hay shoot (which, by the way, was around fifty feet long from top to bottom), the stalls each had one way in and one way out. The possibility of being trapped by an enemy, a fire, or a playmate was great, not to mention the fact that once we were inside the stall, we could not see out. The stalls simply wouldn't do, so we were driven to look elsewhere.

Christian was the first to discover the refuge potential of the hayloft. The hayloft was one enormous room, with nothing to put between our enemies and us. And we were again back to one way in and one way out—which in this case was an old set of stairs. The barn did have a hay door that looked out onto Sweden Street, but it was a good fifteen feet above the ground. Had the barn been burning and the hay door was the only way out, we would have used it, but for ordinary comings-and-goings, the hay door—which offered an excellent view of the entire front yard—was out. My brother had the idea to construct a nest-like platform some twenty feet up in the barn rafters. He made a ladder up the barn wall, at the top of which was a rafter we scaled across to reach his nest in the farthest corner of the barn.

I wasn't comfortable with scaling that distance, so instead of running the risk, I made my own nest a few feet away from the ladder. Soon our friends Tom and Kenny got wind of our hideouts, and they too constructed their own platforms. Tom is my age, so not surprisingly he built his platform across the rafters next to mine; and Kenny, who was my brother's age and more his friend than mine, built his on the far side of the barn, although in the corner opposite of where my brother had built his roost. Our shelters were connected by a series of planks, but we still had to inch our way across several feet of two-by-four to reach them. By the time we were done, the rafters reminded me of the old oak trees that lined our backyard, how each fall they'd drop their leaves, revealing once-hidden bird nests, and the relief and sadness I felt when I realized they were cold and empty and would soon fill with snow.

v. Imaginary Friends

The different locations of our rafter nests illustrate the connection between a child's age and his tendency to seek shelter farther away from the parents' home. Christian and Kenny are only a little over a year older than Tom and me, but the year meant the difference between playing it relatively safe and taking a risk by venturing farther out onto the rafters. Despite this added risk, however (or at least what was from Tom's and my perspective an added risk), insofar as we had built our platforms adjacent to the ladder, and well within reach of some intruder's grasp, Christian and Kenny were ultimately safer in their locations. Clearly, we were at the stage of our development where we had to rely more on the protection of my parents, who were just a scream away, whereas Christian and Kenny—both considerably stronger than Tom and me—were relying more on themselves to avoid any real or perceived danger, including whoever was "It." When they were in their nests, all I could do was look up and scratch my head, anticipating the day when I, too, would be untouchable.

Thus, the distance between our nests had directly to do with our developmental differences. For the first few years of our lives, Christian and Nicole and I played together regularly. But as we grew older (or should I say, as Christian and Nicole grew older), our interests changed and we found ourselves now playing with other, developmentally compatible children. This situation is as old as our species. Given that ancestral children were not only faced with negotiating differences and tensions as they occurred between siblings and other family members, but also as they related to non-family members within and outside the family group, we should expect children to express significant interest in various forms of social play. Play that enables children at the various stages of their development to interact with the opposite sex, as well as to resolve conflict and to cooperate with members of the same sex would seem especially useful given their importance to survival and reproductive success.

I think I would be a very different adult if, as a child, I had to rely solely on the presence of my older siblings or other children before I could begin the long, slow climb toward social literacy. I developed significant attachments to the various pets we had over the course of my childhood, but the most important relationships—one with my dog Gus and the others with several cats—occurred toward the end of my childhood, which is about the time when most children's fear of dogs begins to diminish. My interactions with animals surely played a role in my social development, but not as much as did my relationships with two imaginary friends, one of whom went by the name of Slowdee. Apparently, at one time not too long ago many psychologists—and an all too eager public—viewed children's tendency to interact with imaginary friends as a deficit, as a marker of psychic disturbance. Commenting on this perception, Marjorie Taylor, who in 2004 co-authored a study of one hundred children and their imaginary friends, reported that the parents of one little boy actually came to the clinic with a Bible and, despite having prayed and prayed, were convinced that Satan had possessed their child!

Contrary to the mindset of the past, however, is the emerging awareness that childhood friends are an invaluable and natural part of the child's development. The study found that over two-thirds of children between three and seven reported having imaginary friends, and one-third reported having imaginary friends past the age of seven. Far from being a deficit, then, imaginary friends enable children to develop their sociality without necessarily being in a social situation. I'm sure there are other benefits as well, but insofar as conflict resolution is a crucial aspect of sociality, any play that enables children to develop these skills sooner than later is almost certainly adaptive. For all I know, my relationship with Slowdee was as formative as my relationships with real children and animals, which would come much later. In fact, Slowdee was my first friend. Thanks, Slowdee!

Moreover, the study (which appeared in the journal *Developmental Psychology*) also noted that many children had animals for imaginary friends. Although I suspect the majority of imaginary friends were other children, from an adaptive perspective it makes sense that children should also have nonhuman animals for imaginary friends. On the face of it, such play may appear maladaptive: for what benefit could there possibly be for a child to imagine that he or she is friends with a bear or a wolf or some other dangerous animal? That is, assuming the friend isn't a dog or a cat or some other relatively benign creature? Remembering that children generally form these relationships at a time when they are still very much dependent on the guidance and support of their parents, but are also beginning to express some interest in wandering, having an interest in other animals is part of healthy, normal child development. Learning the ways of other children is indeed important, but so too is having knowledge of other survival-relevant animals in one's environment.

The sooner a child becomes acquainted with the challenges and dangers he or she may face in the future, the greater likelihood he or she will have of meeting those challenges. Whether or not imaginative friends precede physical or actual friends (I am fairly sure that's what happened with me), we can still see how doll play might be considered another important step toward social or "other" awareness. I am not aware of any data that quantifies what percentage of imaginary friends are humans as opposed to other animals, but a glance into Wilder's room (which he shares with a menagerie of animals, including a frog, cow, a puppy, a giraffe, two cats, a monkey, and a mole), or a stroll down the toy aisle of any department store reveals an equal preference for human and non-human animal toys, including dolls and figurines, which have been a part of human life for millennia.

Several factors may affect which animals a child prefers, including the child's age, geographic region, and culture, but generally, we should expect very young children to respond more favorably to animals that would not have posed a threat in the ancestral environment, such as domestic species of animals and wild herbivores. Sometimes Kim and I take Wilder to his grandparents' house to swim in their pool. Wilder doesn't actually swim—I hold him while he kicks and splashes. Thinking he would enjoy floating around in the pool, Kim's father bought Wilder a blow-up crocodile. In contrast to Wilder's teddy bear—which is not even a quarter of his size—this croc, at roughly eight feet long, was practically life-size. If there is one other characteristic that makes it possible for Wilder to interact with his teddy bear, it's the bear's fuzzy, docile, and essentially expressionless face. However, with its enormous, fixed green-blue eyes and menacing mouth brimming with jagged, yellow teeth, the croc's face signified the animal's lethality.

Suffice it to say that Wilder screamed in terror when Kim's father insisted on placing him atop an animal that over the years has consumed its fair share of hominid flesh. In fact, Wilder was so afraid of this croc that he would not relax until Kim's mother had removed it from his sight. Responses like these reveal the explanatory limitations of cultural relativism. I know for a fact that Wilder had had zero experience with crocodiles or any other reptile for that matter. And yet his response to this animal was direct, sustained, and intense. How are we to understand his reaction, except by inquiring into the deep history of our evolutionary past? How can something as ephemeral and temporally insignificant as human culture ever hope to fully explain behaviors that have been millions of years in the making?

The older Wilder gets, the more he will be able to distinguish living animals from imitations. He will begin to prefer animals that provide an opportunity to exercise both his fear-response to and curiosity about potential danger. Children of all ages have at minimum an initial aversion for toy snakes and spiders. In the rare cases that children do have a toy spider or snake, the toys are often used to frighten other children, and so are not played with so much as they are wielded. Much to the consternation of society at large, I was content to play with a doll set originally given to my sister. I also had a soft, silver ape, which, although it was not threatening in any way, did not see much play due to its life-like size. Just the opposite was true with the dolls: at four inches tall, they were easily manipulated into doing my bidding.

When necessary, I would hold all five dolls—three males and two females—between the fingers of each hand and play out a five-doll, group discussion. But usually I focused on the conversations that went on between the alpha male, and his girlfriend, the alpha female. I don't remember what was said, exactly, but conflicts often arose between the alpha male and two other males that rounded out this five-doll band. By far the most significant relationship was between the alpha male and female, the latter representing a kind of foil for the alpha male who was, presumably, the embodiment of the male I longed to be.

The dolls came as a set that included a stage and various instruments, and I distinctly remember playing music by the seventies rock band Kiss as various dramas unfolded on the stage. The most recurrent drama involved the threat of death to the alpha male, which usually occurred when he would tragically fall off the stage right after performing an especially passionate version of the love ballad "Beth." I am fairly sure that the male sometimes did in fact die, but normally he would return to his feet after a few tense moments, during which the alpha female would nurture him back to health. Clearly there was a lot going on during these play sessions. None of this discounts the relevance of things that were going on in my life at the time, but that has not been my point from the outset. While the content of these exchanges could be linked directly to the circumstances of my family life, I am suggesting the possibility that children's interest in using dolls (and imaginary friends, and toy animals) to make sense of human sociality and the environment enhances their ability to survive.

vi. The Dog Days of Summer

One of the advantages of playing with dolls in general and male and female dolls in particular is that I could safely work through the physical, emotional, and sexual tensions that often occur within and between children. As I got older, however, I put away the dolls and became more interested in real children. Interacting with other male children certainly came with its share of challenges, but the majority of these interactions took place through physical contests and so did not really give me an opportunity to develop and test my social savvy or—generally the case with most boys—the lack thereof. Interacting with girls proved just the opposite: one did not usually contest or unite with them in sports or other forms of ritualized mock combat. These contests were crucial to establishing the pecking order among male children, and surely our feats of physical prowess also functioned as displays for the furtive glances of little girls who, like us, had already begun to cultivate their own social skills.

We sometimes played co-ed sports, but those times were rare, and they did not involve the kind of one-on-one interaction so crucial to forming successful future relationships. Just because a boy or girl swings a bat during a game of co-ed softball does not necessarily mean he or she will ever get to first base. Every little bit helps, though. Anything we can do as children to effect better social skills would presumably increase our chances of successful courtship and, eventually, reproduction. In terms of allowing children to explore the differences that confront them— differences that would certainly need to be addressed in order for healthy socialization to occur—some games are more interactive than others.

Although I did not become sexually involved with girls until I was about thirteen or fourteen years old, by the time I had intercourse at age sixteen, I was already acquainted with the ins-and-outs of the female psyche. My mother and sister surely played some role in my exposure to female anatomy. Beyond the implications of the oedipal stage and normal sibling curiosity, these relationships did not represent anything but a superficial introduction to the opposite sex, sort of like a book cover. In any case, like the majority of children I did not view either my mother or my sister in sexual terms. I was naturally motivated to look for hands-on playmates (a curious word in this context) outside my family. I did not have to look far: one of my mother's friends had two daughters about my age who were more than willing to explore their own question with me.

However willing we were to explore our questions together, playing with the opposite sex was unnerving, to say the least. But children are resourceful when it comes to finding ways to pursue their play while minimizing awkward, unfamiliar, or anxious feelings. For us this resourcefulness meant playing Dog, a game that is, given its exploratory nature, similar to Doctor. But at that age we weren't thinking about doctors and would rather have been dogs anyway. That was the whole point of the game: by pretending we were dogs (or any other social animal, for that matter), we could simultaneously explore our interests and curiosities without fear of embarrassment or shame, emotions that can very quickly emerge during play. So mount, sniff, rub, nudge, and bark away! No worries—that's just what dogs do! And to some extent, that's what humans do, too.

As dogs, we were free to explore our nascent sexuality and curiosity about one another while buffering ourselves from any awkwardness. Children have always looked to other animals as wellsprings of information. This information was primarily used to aid children in the hunt, but surely they would have also noticed other aspects of the animal's life, including various social behaviors such as courtship and mating, and the cycles of birth, sex, and death. The child's ecosystem might be considered one big, interrelated, multifaceted and multi-specific schoolroom wherein every child learned what it meant to be human.

vii. A Brief History of Aggression

Each winter snowplows
made the mountain. Thirty boys
divided into kings. And I would stand
between them, waiting for the word

"You!" from whoever was captain; whoever
motioned with the sweep of an arm
that meant I was under their protection
for that hour that day. No one talked

of tomorrow, when the boy who kept you
from falling before would step aside
out of a new loyalty to the ice.

Loyalty to the ice. I've always liked that line. I admit that when I wrote this poem over a decade ago I wasn't thinking about the possibility that aggressive and violent childhood play might be biologically driven. But as I think about the poem now, I realize that it illustrates why aggressive play might have been, and continues to be, important to a child's survival. I was not a violent child; nor was I a target of other children's aggression, which I attribute to several factors, such as having a protective older brother and being a small and relatively quiet boy. I simply did not draw a lot of rival male attention to myself. Given my personality, and the fact I could not physically overpower larger children, I've always been attracted to play that allowed me to experiment with aggression from a distance. There are differing degrees of aggression, and most children very quickly learn to manage their violent tendencies. But aggression and violence still persist in childhood, raising questions about aggressive play's evolutionary significance.

Generally speaking, kids of all sizes benefit from muscular play, but smaller children are at slightly greater risk of mortality than larger and stronger children, despite being no less faced with the need for developing the skills and musculature necessary to survive. We might expect children to play in ways that would promote development and facilitate this outcome. For a small child like me, King of the Mountain offered a rare opportunity to exercise my aggression without necessarily attracting the ire of any particular opponent. Occasionally, personal battles would erupt between individuals, especially the captains of either team, but otherwise the battle was not about particular children. For if they intended to secure the mountain and hold it for the duration of recess, granting them the right to start on top next time (otherwise known as *dibs*), the boys on either side had to work as a team.

Working together and sharing is not an easy thing for children to do—but given the demands of tribal life, and the fact that—like many other social animals—humans are more powerful in numbers, we can see how such play is and would have been tantamount to a child's integration into the group and, therefore, to their longevity. King of the Mountain, and indeed any game that involved team play, gave me and other children of varying size and ability the chance to realize our potential. However, when it came to reaping the social rewards for having held the mountain, including the attention and affections of the girls who watched from a distance, the larger and stronger boys generally took all.

Although this exclusion was painful, I also knew that being the biggest and the strongest boy on the playground was not without its drawbacks. Unlike me and the other smaller boys who simply did not register on the upper tiers of the social hierarchy, the larger boys were constantly battling each other and rivals who sought to replace them. They were targets. But it was precisely their willingness to assume the position of leader or team captain that enabled smaller boys like me to engage in play (and develop important skills) that otherwise would not have been available. Given my size, in one-on-one contests I was disadvantaged, but knowing my teammates were there to support me, I could target even the largest boys, and through that greater resistance, develop my own strength and confidence. Thus, the captains benefited from the status associated with leading and protecting others, and under their protection, the smaller boys were free to fight and dream of taking their place atop the mountain. However much it might seem otherwise, then, in the context of team play every child must necessarily be out for himself.

There is an awfully fine line between looking out for oneself (socially acceptable) and behaving selfishly (socially unacceptable). We like to think teamwork is about self-sacrifice; that our capacity to act on behalf of something "larger" than ourselves is the badge of our nobility. Adults have all sorts of rich and wonderful ways of describing this sense of human greatness. But for a child living in a physical world, and who from early on is confronted with conflict, status, hostility, and aggression (things which will only become more consequential as the child gets older), joining other children for one's self-interests is just common sense. "Common" not because by the age of twelve boys have been steeped in the concept of nobility, but because children who did not express interest in sociality and in learning the ways of the human animal would have found it harder to survive and so would have contributed fewer genes to future generations.

Compared to some of the children I knew, I had it pretty good despite my early struggles. If we have been outside the group even for a short time, we have some basic understanding of why we are predisposed to being social. Late in Sagan's book, the female speaker of *Gangland* articulates the consequences of being outside the group:

One time my sister's kid, he musta got sick or something. All of a sudden he can't move his legs no more. He can't walk. He just drag himself along by his hands. He look real weird. First, folks look away. None of the guys come 'round to babysit no more. Later, they hassle him. Then they attack him. They kill him off, snap his neck. I was sad for my sister.

This is an extreme consequence of being different, but it underscores the general antipathy humans have for difference. Children who were and are motivated to learn the expectations of the group would stand a better chance of benefiting from that association. Although my apparent differences did not invite physical attacks from other children, I was often reminded of my status as an outsider.

Caribou was a small, potato- and broccoli-farming community and attracted few newcomers. That, in and of itself, was enough to draw attention to the new kid in town. Also, my father was a professional in a town mainly comprised of farmers. In terms of how it separated us from the community, my father's job was small potatoes compared to my mother's Mormon background. Actually, a big part of why she agreed to move to Maine in the first place was to escape the rigidity of this sect, especially as her father enforced it. He converted shortly after he met my grandmother, Iris, who was already a member of the church. I guess one could say he quit his life for love. But that kind of love only lasts so long. He was trapped, as though he were walking around in a suit that was three sizes too small. I don't blame him for being angry. But that didn't mean my mom was going to tiptoe around in his cage with him. Her life in Idaho had ended, so when my father was offered a job in Maine, she saw it as a ticket out of there. Trouble is, she was heading to a place that hadn't much seen her kind before.

I don't think my mom ever felt at home in Caribou; it was

as if her father and mother had placed a compass in her hand when she left Idaho, but it wasn't a compass that had any relevance to the world outside the one she had left. They may as well have given her a map to the night sky. But we kids did alright, despite our own struggles with being accepted. When I think back to those years in Maine, I can't help but wonder to what extent my mother's status as a Mormon (a Jack Mormon, really) affected people's willingness to interact with me. It is hard to say. I recall only one occasion when the subject explicitly came up. I stood on the corner waiting for my mom to come out of the post office. The sun was out and the gutters were filled with slush and ice. Some boys I knew from school, two of whom I considered friends, rounded the corner and within a couple of minutes had circled me, chanting something about "little Mormon boy." The refrain was "Oh where, oh where are your horns?" This was the only time I remember being confronted. Perhaps other times some boys just avoided me, or stood silently behind their doors whenever I came calling.

My mother never pushed religion on me, and considering that she herself was already disenchanted with the church, the word "Mormon" was simply not part of my family's lexicon. Consequently, I had no idea what these boys were talking about, but I did know that whatever a Mormon was, it wasn't good from their perspective. That caused me some pain. I'm sure living in that town was much more difficult for people (assuming there were any) who looked different from the locals. Perhaps I shouldn't complain—all I had to do was put on a pair of overalls and I was good to go. If I were a nonwhite, my costume would not have meant anything. Like every other child born into this world, I did what I had to in order to fit in, even if sometimes that meant being violent toward other children who were, like me, just trying to figure out the ways of their kind with as little pain and humiliation as possible.

Out of curiosity, I did some research and learned that in 2000, 96% of Caribou's population of 8,300 was white. I was there from 1968 to 1979, when I'm guessing the population was roughly half the size and 99% white. Not a lot of room for differences in the midst of a human blizzard. Incidentally, when a whiteout occurs, the sky and land disappear in a collision of wind and ice and snow; one's perception—sense of depth, direction, and distance—is lost. Around three centuries ago, a storm blew in from the Atlantic and it has been creating similar conditions ever since.

Sandy, Utah 1979 - 1989

viii. Fight Club

I resented the feeling that I was somehow indebted to larger boys simply because they did not target me for aggression. In fairness, most of the larger boys I knew in childhood were relatively peaceful, as though they understood their power and had enough sense not to exercise it indiscriminately. Better to save their strength for real challengers. As a rule, the bullies were not the largest boys; they were just the meanest and most violent. This complex of responses may explain why it is with relish that I recall the morning some twenty years ago when Eddie Pliskin pummeled another boy who was a notorious bully. I never knew Eddie personally, but I respected him. Not that I really had a choice. We were both about thirteen years old and in the ninth grade, but Eddie was twice my size, easily among the three or four largest boys in middle school. I understood that, as an African American in suburban Utah, Eddie had to fight battles—physical and emotional—every day of his life. Once we entered high school, however, the pool of large boys tripled and so did the potential for conflict.

I never did learn the other kid's name, but I had seen him around and very quickly identified him as someone to avoid. He and Eddie were about the same size, but he was a senior whereas Eddie was a freshman. I don't know what, if anything, this age difference meant to Eddie. Under normal circumstances it would have tipped the scales in the other kid's favor, if only for the reason that, one assumes, he would have had more experience fighting. The kid challenged Eddie just as lunch was getting underway. The lunchroom was an enormous square surrounded by throughways. When sitting at the tables, it felt very much like the audience section of a theater and the throughways seemed like stages—good places to display one's physical prowess.

Eddie didn't say a word. He placed his books in a neat pile on the ground and assumed the stance of a boxer. The other kid smiled nervously and said something I couldn't make out. Eddie dipped and pumped and wound as if his body were a great generator. I'm pretty sure it was at this moment that the other kid realized he was in a world of trouble, and there was no turning back. The kid put up and made small circles with his fists and went at Eddie. Eddie bent his knees and rocked slightly to the left and avoided the kid's first punch. The kid threw a wild left punch. Eddie ducked the blow, and as he stood, his right fist sprang from his body and landed squarely on the kid's jaw, followed by a left upper hook. After the first punch, the kid was already on his way to the floor, so Eddie's second punch really wasn't necessary. Unfortunately for the kid, the one-two punch doesn't discriminate. It's a package deal: you buy one punch, you pay for the second whether you want it or not.

I doubt anyone who witnessed that fight had ever seen anything quite like it. At that age, control and precision when fighting only happened in the movies. But Eddie was a technician. I suspect that his father, knowing Eddie would face more than his share of aggression, saw it as his responsibility to teach his son to defend himself. In any event, that other kid didn't have a chance. Perhaps on some level Eddie saw the fight as an opportunity to get out the word that he wasn't going to tolerate any disrespect. It worked: I don't recall ever hearing of anyone challenging him again. The fight inspired awe and admiration in me and in my friends, who from then on referred to Eddie as Eddie "The Pummeler" Pliskin.

Apart from staving off aggression, Eddie's size didn't seem to result in any added benefits. It's been a long time since I saw him walking through halls on his way to class, but when I picture him he's always alone. Eddie could afford to go solo because of his size and power, whereas the rest of us had to find other strategies for dealing with threats. Having enjoyed, suffered, or simply endured high school, we all remember the different "cliques" or groups of kids that shared the same table at lunch or hung out in certain locker row: the punks, the new wavers, the skaters, the preppies, the jocks, the rockers, the shit kickers. As often as we complain about the existence of these groups, and lament the fact that our children will face the difficulties of finding membership in one, we ourselves belonged to a group—and for good reason. There is safety in numbers.

I benefited from being a skater (including the fact that my skateboard could also double as a weapon), and my membership helped keep me safe. But groups are exclusionary by definition. In my experience, tensions between groups rarely reached the point of physical violence. In fact, that was one of the reasons for forming the group: to discourage attacks from opposing groups. Because this strategy was successful ninety-eight percent of the time, my friends and I had a lot of time on our hands. Recalling the familiar saying "if we don't use it, we lose it," we shouldn't be surprised if groups or gangs devise all sorts of play, contests, and games that allow them to develop, exercise, and retain the skills needed to address any conflict that may occur outside the group.

Although moving to suburban Utah confronted my siblings and me with a whole new set of challenges, for the first time in my life I was not the smallest child. In fact, I was one of the largest and oldest on Huckleberry Court. My brother and I inherited a gang of a dozen white, middle-class kids. Initially, life was relatively peaceful in the kingdom. The longer we lived there, however, the more children we came to know and befriend. Many of these boys were my age or my brother's age—thirteen and fourteen. Eventually, things started to get complicated. There was about a two-week period when our group would get off the bus and walk to my friend Deeg's house. His parents worked well into the afternoon, so his backyard was the perfect place to settle scores by fighting.

Actually, the idea of score settling makes the fighting sound quite a bit more pre-meditated than it really was. Rarely was there a reason for these conflicts, or if there were, it was so inconsequential that any one not involved would not recognize it as a reason. Sometimes the whole day would pass without a combative word and then, just as we were about to go our separate ways, one of us would "call on" another member of the group and off we would go to Deeg's house. As a rule, I did not participate in the fights, even though I was same size as the other combatants. I'm convinced this was because my brother Christian was the most powerful kid on the block.

As the leader of his own gang (known as "The Dead Jelly Beans"), Christian commanded respect from all the kids on our block, none of whom was old enough to be a member of The Dead Jelly Beans. I know that at least part of why I was never invited to "dance" by members of my own group or by members of Christian's group (as well as other kids outside our groups) was because whoever danced with me ran the risk of having to dance with my brother. This was power and status by association, and I know that lots of younger brothers benefited from this relationship. Older brothers were forces, and the more of them there were, the greater force they became, so that dealing with one brother meant dealing with them all.

The Morley brothers—Chris, Pete, and Rob—epitomized this phenomenon. I am certain that every high-school-age kid living in the Sandy and Draper area between 1982 and 1987 had heard of the Morleys. Compared to their collective strength, Christian and I didn't even register. Christian was strong, but the oldest Morley, Rob, was massive, chiseled, and violent. I witnessed Rob's unparalleled brand of violence one night after I drank too much at one of the Morleys' parties. I had wandered outside and stretched out beneath the trees. As I took in the fresh air and collected myself in the darkness, I heard some yelling and saw Rob and another guy standing in the driveway.

A second later, Rob hauled off and punched the guy, knocking him unconscious. The guy went down and would have sunk all the way to the ground if the front tire of Chris Morley's Ford pickup hadn't held him up. Rob grabbed him by the shirt collar and yelled at him. I could see the spit flying out of his mouth as he barked profanities and rhetorical questions at his victim. Apparently, this poor son-of-a-bitch had insulted a girl Rob was courting. And according to Rob's invective, the guy had been warned before getting knocked out.

Thinking back on that night, I wonder how Rob managed to stay out of jail. What Rob did was terrible. We all did terrible things—some I can't even describe. But the bottom line was that having big brothers and big friends was an important part of gaining status, maximizing safety, and securing resources, including females who were attracted to, among other things, displays of power. Even if these allies behaved in unspeakable ways, at the end of the night it was better to have boys like the Morleys on one's side than not. If I were to chart the extent of Christian's and my associations it would reveal a network of a hundred boys. Given our needs to fit in and form allies, I can see how Christian and I, and every other boy we knew, were duplicating behavior that would have been relevant to the survival challenges facing Pleistocene kids and other social animals.

In contrast to public displays of violence, insofar as we were the only audience, the benefits reaped by the victors of our backyard brawls were mostly internal. It is also true that people talk, so although the fight audience was relatively small, winning or losing could still have consequences outside the group. For the most part, however, these fights were about establishing a pecking order and developing skills within the gang. Because these contests were really rehearsals for the real thing, they rarely escalated to the point of significant injury. I think the worst thing to happen was when my friend Chad Herbst kicked our friend Chad Allison in the testicles. Allison did not reappear for a week after that: he would later tell me in confidence that his balls were black and blue. (Interestingly, although Herbst technically was the victor, the group of us scolded him for crossing the line and kicking beneath the belt.) But in general, the fighters were careful not to inflict serious damage. Otherwise the conflicts would go unresolved and the group could fracture and weaken.

We see this same sort of restraint in other animals. For instance, my yard in Arizona boasts a robust population of western fence lizards, and the males use nonviolent means to establish and clarify the herpetological hierarchy. Typically male lizards will sidle up to one another, head to tail, as if they were a pair of slowly closing parentheses. Then, in an effort to show who is bigger than whom, they each inflate their pale orange chins and bright blue abdomens. I have observed this ritual a dozen times in the three years I've lived here, and I can remember only one time when the lizards actually exchanged bites. The two squared off atop my eight-foot, cinder block wall. I'm guessing they were too evenly matched for things to end peacefully, because after sidling for a couple seconds, they lunged at each other and flung themselves off the wall. This dramatic display was exceptional: every other time, the less impressive lizard would turn tail and run.

Animals of all kinds have developed efficient ways of establishing status and resolving conflicts with rivals without seriously injuring themselves. Because early humans obviously did not have hospitals, if they were seriously injured in the ancestral environment, they were as good as dead. Faced with this reality, we should not be surprised that human children seem to have developed similar strategies for safely resolving conflict. Unlike Eddie Pliskin and Rob Morley, each of whom could launch punches without breaking their hands, the gang of boys to which I belonged opted to wrestle. Maybe a couple of punches or slaps were thrown early on, but the moment one of them connected, all pretenses of a disciplined fight dissolved and the wrestling match was on. Wrestling is one of the oldest forms of human combat, but it is by no means exclusive to humans: we see it practiced by other species of mammals and primates as well, particularly by the young.

As we saw with the two Chads, human children and the young of other animals can inflict considerable damage on an opponent. But considering the personal consequences of injury—and that these boys may later belong to the same "team"—inflicting serious damage on one's opponent doesn't make much sense. In addition to the generally restrained manner with which we conducted our contests, we also understood the need to let bygones-be-bygones. Once the melee had ended, the bunch of us—including the often bruised and bloodied combatants—would retire into Deeg's house, where we would raid his fridge whether he liked it or not.

Infighting of some kind is probably universal among children and therefore has a preparative function and importance. In light of our tribal past, children of all ages can be expected to form, test, and practice the skills—physical and otherwise—necessary to successfully compete for resources, defend one's interests, and reproduce. The members of one's own group are instrumental in this regard: without them, individuals would have few opportunities to prepare and thus be less prepared to deal with the challenges that may await them outside the group.

I never witnessed my brother fight the members of his gang, but I would sometimes hear about these fights. For about two years, it seemed he was fighting about once a week. Incidentally, these two years were our first in Utah, a fact which upped the stakes of winning or losing his battles. As newcomers, Christian and I spent our share of time carving out niches for ourselves in the new territory of suburban Utah. For a while, Christian carved with his fists. Had he done otherwise, or had he failed to stand his ground, life would have been very different for me. I owe him a debt of gratitude, for when he fought, he fought for the two of us. Perhaps this is one of the reasons why I was always so nervous when someone would call him out.

Christian only lost one fight as far as I know. I wasn't there, nor did I know the guy who broke Christian's nose, but the loss was personally humiliating. I was bothered by the injury he had suffered, but by that time, Christian was in his early twenties, an age when most people relied on other, less destructive means of conflict resolution. Unlike the fights of past, this was a fight he had more-or-less picked, even though he was drunk and—from the sounds of it—grossly outnumbered. Adult males are powerful and dangerous, especially when they are on their own territory, surrounded by other members of their group. That Christian didn't see this coming attests to the extent of his intoxication, but impaired or not, I felt a profound sense of embarrassment, hurt, anger, and shame.

Despite the gravity of this response, what interests me more is that I would have it in the first place, that is, since so much of what I felt had to do with how his loss would affect me and my reputation. It was as though that single loss had undone everything he had built up to that time. I know it sounds selfish: my brother gets beat up and all I can think about is myself. I never expressed my disappointment to Christian, and I didn't always worry about how the outcomes of his fights would affect me. Or if I did, it wasn't obvious. On a deeper level, however, when it comes to being part of a social group in which one's own status and treatment is affected by the behavior of other members, guarding one's reputation is just good planning. We want our reputation to precede us.

I remember riding home on the school bus and seeing a couple hundred kids swarm to the church parking lot next to school to watch a fight. It was a bizarre but common sight. That school officials rarely learned of these spectacles until they got phone calls from angry parents suggests a profound disconnect: usually when a fight was going to take place, other children would broadcast the event all throughout the day. Back then, fighting was treated as a rite-of-passage, so it was not unusual for some teachers (especially gym teachers) to turn a blind eye to bullying and skirmishes between boys, even though they very often would escalate into black eyes, bloody lips, and knocked-out teeth. Mostly I worried that one of those boys in the midst of that bloodthirsty throng was my brother.

Not being a fighter myself, I hated it when Christian fought. I feared that he might be hurt, but I also knew, perhaps unconsciously, that if he lost our joint status in the neighborhood would be jeopardized. Long before Christian had his nose broken in a drunken brawl, I watched him trounce a boy twice his size. I don't recall what the fight was over, but this kid had been the dominant boy in the neighborhood prior to my brother's arrival. Under the circumstances, that was the only reason this kid needed. After school, four or five of us got off the bus and made the long walk in relative silence. It was odd, all of us walking together. I wasn't even the one about to fight and yet my adrenaline was surging. I remember walking next to my brother and looking at him. He didn't look like he was about to fight. His face was smooth and calm and faintly rosy from the November wind blowing off the high desert and into the encroaching suburbs.

Once we reached the selected site—the front yard of a vacant house—the other kid started talking trash, rolling his head, and throwing air punches. It was cold that day, but instead of warming up, my brother stood there and waited, his sneakers firmly planted in the crusty snow. The kid was about a foot taller than my brother and was rumored to know kung fu—which meant he had a pair of nun chucks made from an old broom handle and had watched at least one Bruce Lee movie. The fight broke out. The boy kicked like he was on stilts, but he managed to land one on Christian's side, knocking him off balance. The kid charged in, fists flying like pale, rubber hammers. This guy was no Eddie Pliskin, however, so instead of pulling back and regrouping, he kept on going, until finally he was within my brother's reach.

Christian grabbed him and wrestled him to ground, where he punched him in the side of the face. When it was clear that the kid was beaten, Christian asked him if he "gave." Through a soup of snow and spit and blood, the kid said he was going to whoop my brother's ass. Christian took the kid's head in his hands and pounded it into the snow. A couple of moments of that punishment, and the kid conceded. Christian stood and backed away. Apart from a torn sweatshirt and two big wets spots on his knees, he was unscathed. My worry instantly transformed into relief and mild elation. But I noticed the sudden hardness in my brother's young face and the blood in the beaten snow, and I felt like whatever was kind and gentle in the world had taken one look at us and flown away.

ix. Bloodletting and the Damnedest Things

As a kid, I was mesmerized by Bruce Lee. My mother bought me a book about him and I remember enjoying the large photographs of Bruce in various fighting stances. Perhaps one of the most famous shows a glistening, muscle-rippling Bruce looking directly at the camera. His fists are up and he has a calm but defiant expression on his face. Although he looks perfectly composed, his cheek and chest appear to have been raked by a three-pronged weapon of some sort. The folks in make-up did a fine job of creating a welling effect—the wounds appear on the brink of spilling blood down his face and chest. Emblems of a warrior who has seen and survived battle, the scars gave Bruce a certain sex appeal, much like the slash of scar tissue on the old G.I. Joe doll's face. But Bruce was no doll. He was living proof that power and charisma come in small packages.

My childhood dreams of learning to fight like Bruce Lee were short-lived, but even now it is hard not to be awed by him. Sometimes I'll come across one of his movies at the video store and sheepishly present it to Kim like a child who pulls on his mother's shirt and asks if she'll buy him a toy. Kim is remarkably patient and agrees to rent the video—provided we get one other as well, preferably something with Johnny Depp. I happily acquiesce. I figure she can fantasize about Johnny while I'm busy whooping Johnny's ass, Bruce Lee style. My all-time favorite fighting scene is at the beginning of *Fists of Fury*. Bruce plays a long-lost nephew who takes a job working as a waiter for his uncle. The uncle has fallen on hard times because his restaurant is being "shaken down" by some local thugs. Bruce hasn't taken his first order when the thugs appear, demanding payment.

These thugs are of the courteous variety. Although the uncle, emboldened by his confident young nephew, refuses to pay them their so-called protection money, they agree to take the fight to the back alley lest they trash the place. Once outside the thugs spread into a half circle, weapons in hand, as Bruce gestures to fellow restaurant workers to give him some space and, in his best-dubbed English, tells them not to worry. Bruce drifts down the line of thugs, carefully looking over each one of them before returning to his original position. If there is any doubt in the minds of the thugs that they are dealing with a master, it should disappear the moment Bruce begins his warm-up. It begins with simple but psychologically unsettling knuckle cracking and head rolling, and culminates in a flurry of kicks and punches, the speed of which is marked by the sound and the blurring of his clothes as they snap across his body. This warm-up takes place within a matter of seconds, so when it's all over and Bruce stands at the ready, beckoning the thugs with his outstretched hand, the thugs and I are asking the same question: What the hell just happened?

Perhaps one of the things that makes this display so effective is its ability to mesmerize the opponent, who will—if only momentarily—be vulnerable. If there is a continuum of fighting strategies ranging from basic to sophisticated, this surely constitutes an example of the latter. Bruce also relies on defensive and offensive strategies that are not only basic to humans, but to other animals. Again, the lizards with which I share my yard come to mind. Keeping in mind that animals would rather not risk injury by fighting, the ritual of battle includes an assessment and/or withdrawal period in which a combatant may opt out of the conflict if he decides that he is outmatched.

How male lizards behave when in conflict is revealing in this regard. First of all, unlike some presidents of the United States, male lizards do not indiscriminately rush into battle. Instead, the moment they become aware of each other, they begin their displays. These displays resemble push-ups, and once the lizard is elevated, it inflates its body from chin to vent. Lizards have relatively good eyesight given that they will display even when they are several feet apart. Because these long-distance displays give opponents that much more time to back out before things get ugly, they support the claim that animals (except some humans, of course) use violence as a last resort. When neither lizard backs down, the two rivals begin to close the distance between them while increasing the frequency and intensity of their displays, almost as if they are saying, "Ok, I'm serious. See?" (Display.) The other lizard says, "Yes, but I am really serious. See?" (Display. Then a couple steps closer.) "You better stop right there. I'm really *really* serious!" Most of the time, this strategy works and one of the lizards retreats. But if this doesn't work, the lizards cross into a space where physical conflict can occur.

What fascinates me about this stage of the conflict (and what reminds me of Bruce Lee) is how the lizards position themselves. Perhaps in a last ditch effort to display their might, when they are within inches of each other the lizards inflate and sidle up to one another, head to tail, until they have formed a broken circle. Judging by how their bodies quiver, the lizards must expend considerable energy to sustain that level of elevation and inflation. Perhaps this final display of staying power represents the last chance the opponents have to flee, although if one were to do so he would surely incur a bite to the tail. This is the importance of the head-to-tail position—if, during the early stages of the conflict, the lizards were to charge each other head-on, they would not be able to display their size and color and prevent a physical confrontation. But even in the worst case scenario when neither lizard backs down and attack seems inevitable, the head-to-tail position functions as a buffer against lethal damage by putting distance between the heads (or the brains and eyes) and the mouths (or the teeth) of the opponents.

This is not to say that fence lizards and other animals will not inflict serious damage to their opponents if given the chance; under the right circumstances, they will kill if they can. But I wish to point out how animals, despite their differences, have evolved parallel strategies for dealing with physical conflict, including their physical orientation to the threat. Lizard or human, the goal is to keep oneself away from the business end of one's opponent while remaining within striking distance should the opportunity to strike present itself. A male western fence lizard and Bruce Lee achieve this goal differently, each according to their intelligence, anatomical limitations, and locomotion.

A cinematic example of how some species augment this distance occurs in Stanley Kubrick's classic film *2001: A Space Odyssey*. The opening scene reveals a group of early primates (presumably *Australopithecus africanus*), one of which finds a large femur bone and taps it on the ground next to the skull of an herbivore, possibly a rival. The more he taps, the more excited he becomes, until finally he crushes the skull into hundreds of pieces, a feat that would not have been possible using his bare hands. Whatever questions we might have about the meaning of the scene, we recognize the bone-wielder's violent self-awareness and the implicit warning that things are going to change around here. Apart from enabling the weapon wielder to crush the skulls of enemies and prey species, another thing that made the leg bone so effective is that it added an additional two feet to the wielder's reach.

Thus, the weapon would have been all-important in terms of creating distance between one's self and one's target. We can see the appeal of projectiles such as rocks, which could be thrown from an even greater distance and with potentially devastating effect. My son Wilder already exhibits an interest in throwing things, especially objects that fit firmly in his hands, such as his little wooden blocks. Watching him, I wonder if throwing isn't among the first motions that children learn. Children seem to take pleasure in grasping a smooth stone in their hand, selecting a target, and testing their aim. The projectile itself is arbitrary, because children throw whatever is available. (When we were kids living in Maine, my brother and I had some pretty wicked toy fights at night in our bedroom.)

As I noted earlier, my first rock fight was my last. It simply wasn't practical. I can still see the rock coming at me, and I remember turning to escape it, only to be struck in the back of the head. I guess I shouldn't complain, for if I hadn't turned, the rock may very well have hit me in the face. My brother's friend Kenny had thrown the rock, and once he heard me wailing and saw me heading toward his house, he and my brother made a beeline to my side and tried to console me. But to no avail—by then I had put my hand to my head and had seen blood, and in my child's mind, there was nothing worse. Within moments I would eagerly reveal his misdeed to his mother, Zelma, who was French, fierce, and always wore pants.

In terms of sophistication, stones and bones are fairly comparable, but given the likelihood that stones were more abundant suggests that they preceded bones as weapons and as tools. In the context of my own experience, rocks were the first in a series of weapons I would use to play-fight my friends as well as to hunt and kill other animals. By the time I had reached the age of twelve and was living on Huckleberry Court in Sandy, Utah my interest in rocks had waned, replaced with an intense attraction to spears. It was an interest I shared with every boy I knew. When it came time to patrol and hunt in the desert or the woods, no one ever needed convincing. We were all little killers-in-waiting.

This is not to say that we were equally deadly, just that we all had an interest in hunting, stalking, and, at times, administering the *coup de grace*. My friend Nole Walkingshaw was one of the deadliest boys I knew, but he was also among the most gentle. I called him just the other day to ask if he remembered the time we speared to death a prairie dog at his summer cabin outside of Heber, Utah. That was twenty-five years ago, so I wasn't expecting much. But he answered my question so quickly it was as if I had asked him about something he had done yesterday. "Oh yeah," he said. "I still see the blood on my spear point." After a long pause, Nole mused, "We did the damnedest things back then, didn't we?"

Nole pointed out that after we had killed the prairie dog, the tone of the day changed, as if we were suddenly beside ourselves and compelled to reflect on the meaning of what we had just done. Although I cannot verify his memory, he also thought we buried the prairie dog "to hide the evidence," which returns us to the eco-moral implications of killing for non-survival relevant or wasteful reasons. Whether or not we acted on this awareness, I think most of the boys I knew recognized that needlessly killing other creatures was wrong. The older a child gets, the more culture influences the nature of this recognition, but culture doesn't cause morality as much as it shapes and expresses it. Although we did not understand our feelings of guilt and the naturalized moral awareness that prompted them, we abandoned our spears and instead took up arms against each other.

One of the advantages to stick fighting was that we could give free reign to our hunter instincts without seriously injuring anyone or killing other animals. In contrast to the stick fighting I did in Maine, stick fighting in Utah was considerably more involved and intense—we were older and more powerful. At thirteen and fourteen years of age, my friends and I began wandering farther and farther away from home, which imbued our play with a sense of uneasiness and danger. My friend Deeg had lived in Utah for a few years prior to my family's arrival, so he had already done some wandering and found some rich play areas. The most important place was Bell Canyon Reservoir and the surrounding area. That my friends and I would walk three or four miles to get to the trailhead is testimony to how much we enjoyed the "rez," but the walk wasn't the only obstacle to our enjoyment. Each time we made the trek, we ran the risk of being captured by the "reservoir man," a tall, white-haired old man who saw it as his duty to guard the trailhead with his white German shepherd.

I am used to swimming in fresh water and in natural settings, so the reservoir was the first place I thought of each summer when the temperature started rising. I could not care less that I was swimming in peoples' drinking water. "To hell with them for building up here!" was my general attitude. The reservoir man anticipated trespassers like me each summer. I know we anticipated him. In the early days, before we discovered another route up to the reservoir, we had to pass within a couple hundred yards of the reservoir man's home. This was always a tense time. On more than one occasion, we got chased out of there without having so much as dipped our toes in the water. After some practice, however, we all knew the drill: no talking, stay low, and move quickly. This usually worked, but the reservoir man was smart. We could do everything right and still something would go wrong.

One of the last times I saw him, the reservoir man must have been sitting in front of his living room window because we had no sooner crossed his property line when he appeared with his dog, and the four of us high-tailed it out of there. A few minutes later we regrouped and walked back up the hill to try another approach. The old man was crouched behind sagebrush. He leapt out, and the chase was on. Out of the corner of my eye I saw him chasing after my friend Curt, who, unlike the rest of us, had very short legs. I guess the old man figured that made catching Curt a sure thing, but he didn't count on the nitric-oxide effects of fear, which translated into a burst of speed and Curt's complete willingness to crash through the trees and low branches, where the reservoir man could not easily follow.

Although the old guy was something of an inconvenience, we recognized him as a worthy adversary. Many years later, I ran into him again at a yard sale he was having on his several-acre lot. Over the years, he had accumulated several Volkswagen Beetles, a jet engine, and several other large machines that were strewn across his property. Some had been there so long that the vegetation had grown over so that they looked like woolly animals or burial mounds. He had to sell it all because the State of Utah, by virtue of imminent domain, was about to build a road through his property. The land surveyors had already staked out the path the road would take. I didn't see his dog. A tough old sumbuck, the old man had probably outlived him. I fantasized about revealing my identity: one of the little shits who used to steal across his property on my way to the reservoir. Despite being only nineteen, I did know one thing: a man in a state of loss should not be asked to revisit what may or may not have been a happy period in his life.

By the time I entered my first year of high school, Chad Herbst's parents had bought one of the mountain houses whose drinking water we apparently sullied each time we swam in the reservoir. After school, assuming we went to school at all that day (high school could not compete with the mountains), we would meet at Chad's and hike the path behind his house. Usually we knew well ahead of time that our primary purpose for going to the reservoir was to stick fight. Perhaps someone had mentioned it on the trail, or maybe one of us picked up a sword-sized stick and held on to it despite the fact that hiking would be much easier without it. It was not unusual for one of us to launch an attack, hurling a piece of wood just as we were nearing the reservoir. But typically we were very methodical and performed certain rituals prior to commencing the stick fight.

Though we functioned as a group, each of us preferred the company of some members to others. This made deciding teams fairly easy. If I still knew these boys, I would be interested in hearing their interpretation of our relationships and individual positions within the group. I don't think they'd disagree with me when I say that I got along well with most everyone. So for me, then, the teams were more arbitrary than anything else. This wasn't the case for everyone, though. I am sure that at least a couple of my friends chose their teammates not on the basis of whom they wanted to fight with, but on the basis of whom they wanted to fight against. Perhaps that was another function of the activity: it provided certain boundaries within which we could resolve tensions while minimizing the risk of group disintegration.

I know there were a couple of times when the fighting turned ugly, but usually the game ran its course without incident. Apart from some minor taunting, we were usually willing to set aside our grudges for the sake of play. After reaching the reservoir and scanning the area for official-looking people and vehicles, we would walk the boulder-strewn shoreline to the inlet on the far side of the reservoir. By summer the stream's flow was clear, low, and steady, but it had raged all spring long and deposited tons of rock, dirt, and silt, forming a shoal that extended fifty yards out into the water like the tawny backbone of some mythic lion.

Throwing our sticks aside, we would race the length of the shoal and then out into the mountain water. When the water reached our waists, we dove and swam to the center of the reservoir, where we laughed and cursed the cold on our sunburns and agreed there was no finer thing. Back on shore, we found a silky, sweet-smelling mud infused with flecks of pyrite and granite, which we would lavish on our bodies and carefully apply to our faces. One of my closest relationships was with the sun, so it was already difficult to detect my sun-browned body in the trees above the reservoir. But in my stripes and spots of mud, I became another encrypted forest creature, a breathing shadow.

Armed with sticks and dressed for war, my teammate and I would depart the inlet and head into the woods above the reservoir with the understanding that the other team would soon pursue us. Considering that the theoretical purpose of stick fighting was to fight with sticks, it may seem odd that we rarely came to blows. If by fighting we were attempting— among other things—to (re) establish a pecking order within the group, it might seem as if the activity had no purpose at all. If we did not come to blows, how could an opponent be bested and put in his proper place? If we have ever watched a game of basketball or any other non-contact sport, we know there are lots of ways of being bested other than losing a hand-to-hand physical confrontation. For instance, there is nothing worse than being "dunked on" in basketball or "struck out" in baseball.

Similarly, although hiding doesn't get nearly as much attention and praise as running fast, throwing far, jumping high, and fighting well, the ability to hide well is just as— if not more—important as any of these displays of physical prowess. I don't think it's an accident that hide-and-go-seek is significantly more popular among young children than among older children. Compared to older children, young children are quite limited in terms of what they can do physically. These limitations make children vulnerable to predators and ill-intentioned humans, suggesting why kids have an interest in skillful hiding.

Hiding is not the only or even the most important conflict-avoidance tactic of young children. Kim, Wilder, and I live across the street from an elementary school, and as I sit here writing with my little cat Bella Jean lying in the open window, I hear the shrill screams of children at recess, just as I have for the last three years. I know it's socially convenient to suppose that only little girls scream, but as anybody knows who has spent time around children, little boys are just as eager to scream as their sexually dimorphic counterparts.

In fact, children of both genders start practicing and experimenting with screaming right around the time they start playing hide-and-go-seek and other developmentally appropriate, survival-tailored games. Whether or not screaming will serve an imperiled child (or a fledgling mockingbird) in eliciting the aid of a family member depends largely on proximity. But in rare cases where the child is alone, screaming would, among other undesirables, alert a predator to the child's location. Hiding is perhaps a child's last—and arguably most dire—resort. Far from being an insignificant form of play, hiding may be natural selection's gift to a vulnerable animal.

As thirteen-year-olds, my friends and I developed aggressive abilities such as throwing, running, fighting, and stalking, but even so, we spent the first half-hour of our stick fights indulging our ancient attraction to hiding. Hiding well wasn't the only way to display our woodsy prowess. We also valued stealth and the ability to stalk well, the finest expression of which was to surprise an opponent. The human psyche makes it easy for us to believe we are somehow as beautiful, strong, smart, sexy, lethal, or desirable as the next person. This innate sense of vanity emerges when we are very young and diminishes after our prime reproductive years. Thus, for me at least, the sting of being bested at thirteen or fourteen years old was significantly more painful than it was in my twenties and thirties.

Although I didn't consummate my sexual potential until I was sixteen, like other boys I knew at the time, I had already begun to experiment with sex years earlier. I wonder if it was always so humiliating to be bested as a child because there was more at stake. If a boy gets a reputation of being a "wimp" or "sissy" at the age of thirteen, it may have serious consequences in terms of how he is later perceived by potential mates and members of his social group. Humiliation is never good for one's reputation, but from the perspective of how it would affect one's chances of attracting a mate, being outdone in one's twenties or thirties seems a little after the fact. I think most people would agree that, primarily from a social perspective, the early teens are some of the most difficult and socially formative years of a person's life. This might help to explain why I remember with such clarity the time I was surprised (and therefore bested) by my friend Paul Hanzelka.

However enjoyable it was to prowl those woods above the reservoir, eventually we all gravitated toward "Syd Rock," a large granite boulder that had sheared off the canyon walls thousands of years ago and had the dubious distinction of bearing the name of my brother's girlfriend. Syd Rock rose in the midst of a boulder field. The area lent itself to play because it had natural boundaries—we did not expect to be engaged outside the field. In any event, I was well within the field—perhaps too far within it—when I wandered into one of the many small caves between boulders. I don't recall whether I was looking for a hiding place or the enemy or simply exploring, but after I had wandered in about ten feet, my friend and opponent Paul suddenly appeared in the entryway.

He had been waiting high atop the boulder in anticipation of this blunder; the moment he saw me enter the cave, he down-climbed the near vertical boulder face some fifteen feet and then dropped ten feet to the ground, landing in front of the entrance to the cave, simultaneously scaring the bejesus out of me and blocking my escape. What strikes me now is how clearly I can see Paul standing there. Considering he had just cornered and bested me without so much as raising his stick, his face was remarkably calm and composed. He didn't gloat. Instead, he raised his eyebrows and waited for me to drop my sticks in the dirt and walk out. Although I felt the sting of humiliation, my discomfort was minimal compared to what others had suffered from their run-ins with Paul.

Paul and I were pretty good friends, so after I had surrendered we sat down and smoked a cigarette and waited for the others to join us. Even if we hadn't been good friends, hard as it would have been, I still would have deferred to him. Paul wasn't much bigger than me, but he was very strong, fearless, and aggressive. I was none of those things, so it was easy for me to lay down my weapons rather than risk a blow and compound my humiliation. Some of us weren't as quick to respond to Paul's dominance. I recall the day Deeg felt Paul's wrath because he didn't disarm quickly enough. Deeg had lowered his stick, but he had not dropped it. The skirmish happened as we were preparing to leave the reservoir and make the long walk home, so I was a little annoyed when Deeg refused to promptly concede. The two argued as we made our way down the trail, until finally Paul struck Deeg's stick, violently knocking it out of his hands.

Deeg protested by calling Paul every name in the book. Usually Deeg's English accent was subtle, but whenever he got upset his accent became very pronounced, almost to the point of unintelligibility. Paul was somewhat amused by Deeg's tantrum, so rather than hitting him with his stick, Paul strong-armed him and almost succeeded in pushing him down a steep embankment. That was one of the things that made Paul so intimidating: I didn't think for a second that just because he was laughing he wasn't going to knock Deeg on his ass. Had Deeg not had a foot firmly planted beneath him on the embankment, he would have gone for quite a ride. Eventually he realized his predicament and fully conceded. By then he had been stripped of his weapons, so all that remained was his pride. But when Deeg realized that Paul was still more than willing to give him a good thrashing, pride gave way to wounded deferment.

The face is remarkably communicative in that way: the slightly down-turned mouth, rumpled chin, and downcast eyes, are unmistakable signs of a forlorn and submissive animal. In *The Descent of Man*, Darwin ponders the "unbroken inheritance from a common progenitor" by pointing to the expressive similarities between human faces and the faces of other primates such as gorillas. Might this notion extend to non-primate animals as well? A few months ago, PBS aired a program about the reintroduction of wolves into the Greater Yellowstone Ecosystem. One scene was particularly effective in addressing the impact wolves are having on the coyote population, but also how utterly recognizable are certain facial expressions.

Big flakes of snow fall on the hills and on the carcass of an elk the wolves had killed. A beautiful, full-coated coyote enters the frame and begins tearing away tufts of fur until reaching the elk's hindquarters. The camera cuts to an exquisite black wolf whose hackle appears to be tipped with ice. The wolf does not share the frame with the coyote, but we know he has seen the coyote feeding on the kill; in an instant the wolf's ears lie back, his neck and head project out and down, and his entire body lowers as he races through the snow toward the coyote. The pack observes his behavior and soon joins him. They are in full agreement. The camera cuts back to the coyote just as the black wolf rushes in and attacks. The wolf's speed is so great he cannot adjust when the coyote shifts to avoid the lethal jaws. In that second after the miss and before the other pack members rip him apart, the coyote tucks its tail between its legs and cowers. The expression on the coyote's face is unnerving. I had never seen such an encounter before, but it was all so painfully familiar I felt as if I were remembering it.

Had the coyote made this gesture in the context of its own social group, it's unlikely that any violence would have ensued. However, wolves are not receptive to displays of submission from an animal it would be better off killing. This is what made the event so memorable and disturbing: despite its unequivocal show of surrender, the coyote was violently dispatched. Given the coyote and wolf's antagonistic relationship, there is no reason why things should have turned out differently. But from my perspective as a human animal, the display was effective in eliciting empathy—in the context of that emotion, I experienced the coyote's death. There was a moment when I felt the wolves would relent and the coyote would survive. However unfounded my expectations were in this case, the one response is built into the other. Empathy between humans is a powerful adhesive: it can cause one human to help another. When the opposite outcome occurs—even if it occurs between a coyote and a wolf—we are thrust into a state of distress. We are simply unprepared for this ancient recourse to fail.

However uncomfortable I may be when witnessing this

violent drama play out between other species, most of us have much greater difficulty when the violence transpires between humans subsequent to a show of vulnerability, submission, or surrender. Steven Spielberg's 1998 film *Saving Private Ryan* offers an interesting example. At the end of the film there is a fight scene between an American and a German soldier. After a few moments of intense melee, the German soldier overpowers the American and wrestles him to the floor. Lying atop him, he sinks his knife slowly, almost lovingly, into the American soldier's heart.

Just before the stabbing, the American surrendered and pleaded, even as his killer was sliding the blade into his chest, stroking his hair, and whispering "Shhhhh" into his dying ear. The scene was a poignant reminder that, beneath our civility and sense of cosmic specialness, unspeakable violence is alive and well within the human species. Perhaps this is why war films (and not actual wars) are so popular: they reveal the limits of our definitions of humanity. Suddenly the relationship between the wolf and the coyote does not seem so remote and brutal. The recognition that, as humans, we share and are driven by many of the same needs as other animals might seem like a convenient way to excuse or justify harmful human behavior. In reality, however, one of the primary outcomes of understanding the biological underpinnings of human nature is the *prevention* of damaging behavior, regardless of whether it is toward other humans or the nonhuman environment.

To suggest otherwise is to adopt the indefensible position that the more we understand human behavior, the more we can justify it. But biology has nothing to do with justice, at least not in the way we normally understand the concept. If anything, as we gain a better grasp of why we behave in the ways we do, we will become increasingly effective at preventing, addressing, and treating undesirable behavior in our species. Given this largely untapped understanding, it is truly alarming and baffling to note the hostility and suspicion many people have toward science in general and Darwinian evolution in particular, unless of course we are talking about the "science" of making good shampoo and other consumer products. We seem to value science as long as it is inconsequential, impersonal, and held at arm's length.

x. Hunters and Drivers

Slashing and killing a man with a sword
offered visceral pleasures not found in guns.

—Barbara Holland

Humans would have been (and still are) the greatest mortal threat to other humans. Children ought to be predisposed to play games that involve pitting themselves against other children, familiarizing themselves with the habits of our most formidable adversary. Incidentally, this tendency may be linked to the popularity of "shoot-em-up" video games, especially those which pit players against other players or computer-generated human foes. These violent games are attractive to pre-teen boys, but their primary audience is the thirteen to twenty-four age group—the group most preoccupied with competing for status, mates, and so on. Although the games offer vicarious experience, they tap into the primal, competitive preoccupations of the male psyche. This fact no doubt informed the U.S. Military's decision to develop *America's Army*, a $16 million video game used to attract future soldiers and recruit young men. What form the contest takes in real life, however, depends on the child's age and development.

Whether it was by running faster, stalking better, throwing farther, out-hiding and, occasionally, out-fighting—all within the context of the natural world and the challenges it presented—stick fighting (as well as the other games we played over the course of our childhoods) linked us to past challenges and prepared us to live a life that we have not seen for thousands of years. But as the familiar fight-or-flight response implies, what has not changed is that most of us use violence as a last resort. The adaptive importance of any childhood game lies in its ability to effect development, impart necessary survival skills, and to help us learn to resolve conflicts non-lethally. Thus, our games rarely entailed bloodshed, but they would not have met their adaptive potential if they did not also prepare us for worst case or violent scenarios.

Note, however, that violence during Pleistocene times was very different from today, and not because people today are necessarily more violent. Rather, modern humans are without question more lethal than they were in the past, largely because of "advances" in weapons technology. Our ancestors had good reasons for avoiding violence, one of which was the fact that they would have been personally involved. In their case, the avoidance of violence was linked mentally and physically to the likelihood that they would suffer physical injury. Even today this must still be true to some extent. But when I look at my personal history and think about the billions of dollars spent on developing ever-more sophisticated and long-distance ways to destroy other humans, it seems our relationship to violence has become much like our relationship to science and nature: impersonal and remote. Perhaps this is how some of us can profess our support for war as a means of resolving conflict (I saw a bumper sticker the other day that read "Give War a Chance"), but not have the stomach to view war's actual destruction.

In a way, each child represents an unfolding, living history of survival-enhancing strategies, and sometimes these strategies have nothing to do with doing the "right" thing. By the time I was fourteen, my interest in warfare had burgeoned to include BB guns. BB guns had been part of my life for many years, but until then I had never considered using them on other children. The fact that there were so many of us and so few BB guns suggests that a significant percentage of our parents knew better than to arm their children. In any event, I'm sure that my mother and the other parents who succumbed to their child's pleadings to buy them a BB gun had no idea we would draw a bead, as it were, on other children. In retrospect, I wonder how they could have possibly thought we would do otherwise.

I should qualify this last remark by saying that unlike a few of my friends who hunted with their fathers as children and today hunt with their own sons, I did not grow up in a gun family. I must concede the possibility (not the certainty) that I wouldn't have hunted my playmates had I been explicitly warned of what might happen if I did. This sounds totally insane. How, as a fourteen-year-old boy, could I not have known the potential for disaster? The truth is, I did know—to an extent. I knew that shooting BB guns at other children was dangerous in the same way that most sixteen-year-olds know that driving a car too fast is dangerous. So why wasn't that knowledge enough to prevent me from shooting at my friends, or to keep many teenagers from having lead feet?

Again, the answer seems founded on our biology, which, although it may be offset through various forms of education, has prepared us for some things and not for others. In the absence of this education, and even sometimes in the presence of it, teenagers may not grasp the gravity of shooting BB guns at each other or driving too fast down that canyon road. The threat posed by these modern dangers just doesn't fully register. John Alcock sent me a pair of juxtaposed images that I think may speak to the biological part of the answer. On the top left corner is a very memorable photograph of a man being bitten in the face by a large snake. The other image appears on the bottom right corner of the page, this time of a mangled heap of two wrecked automobiles that had collided in a head-on accident. Alcock shows the images to his sociobiology class to help illustrate the adaptive significance of our fear responses, and how we tend to be significantly more fearful of dangers from our evolutionary past (the snake), despite the fact that modern dangers (like car accidents) are far more common and lethal.

Gordon Orians has also gathered considerable research that shows how people generally respond more aversively to images of snakes and spiders than they do to images of guns, car accidents, or frayed wires, even though we are much more likely to be injured or killed by these modern threats than we are to be dispatched by a snake or spider. According to Orians, "No purely cultural/learned interpretation of phobias can explain these striking results. The nature of the responses makes adaptive sense. It pays to retain awareness and caution with respect to potentially dangerous situations and objects." I wonder, then, if this helps to explain why we so often behave maladaptively or, in my case, shoot BB guns at each other.

On the timeline of human hazards, snakes, spiders, precipices, deep water, and lightning are primordial. Guns, however, are entirely recent. In addition to the temporal chasm between the Pleistocene mind and the gun, the latter also kills with a speed and ferocity that is hard for most of us to imagine. Something about it just doesn't compute, except for those who have experienced it, and even they often say there are not words to describe what they witnessed. As creatures who evolved during Pleistocene times, most of us are simply not prepared, and do not have the psychological vocabulary to process the meaning and impact of modern technology, including the seemingly innocuous BB gun.

It is hard to understand how humans can create things that are so removed from our common understanding, but that is exactly what is happening. The more technologically oriented we become, the more pressing, difficult, and necessary it will be to educate our children about the dangers and responsibilities involved in using and living with those technologies. A well-rounded and relevant education would include a discussion of our evolutionary origins. Without a clear understanding of the depths of human nature and our membership in the community of common descent, we cannot expect to use technology responsibly. Hunters' and drivers' education be damned.

xi. Denouement

I've seen a boy shoot a hummingbird point-blank as it sipped from a feeder. I could not look at him for a long time after that, and never in the same way. I heard from a friend that he now holds a Ph.D. in fishery biology. Though that seems contradictory, it is a familiar story—I, too, changed from an indiscriminate killer of the wild to a killer with an environmental ethic. Like millions of children who've preceded me, one of the ways I learned the importance of life was by taking it. I know how bad hunting and killing can feel, and how good. Understanding how I can have such different feelings about the same event has been crucial to my own sense of well-being, but now I think about how that knowledge might be useful to my son as he grows up. I need to start telling the truth about my life so I'll be prepared when Wilder starts asking questions and expressing his own tendency to hunt and, perhaps, to kill. I will not encourage him to fear his nature. Instead, I'll tell him stories that teach him to embrace, celebrate, and—when necessary—temper it.

Sometimes when I'm fly fishing and I've got to cross the river, I'll throw out caution, enter the current, and let the river take me. I just go with it. But the conditions have to be right. I need a warm summer day to float-walk and chance a fall in the cold mountain water. I've got to know there's no harm in it. In winter, however, the sun is no longer on the river bottom and the water runs high and black. The world is cast in ice and snow. Instead of free-footing it, I use a sturdy stick to help me over the green stones and through the stretches of fast or deep water. There will be these times and times when the stick doesn't touch. But we can use our stories to help our children find their footing in the world and judge the depth of their nature.

I could tell Wilder about the last time I hunted with a gun and how a single experience can change us long before we understand it. Usually we feel before we think. Assuming we think at all about our feelings, a kind of delay occurs. In my case, the delay between feeling and meaning was about twenty years. I was still living in the house on Huckleberry Court, but things had gone south for my mother and my stepfather Phil. It was hard to stay put and watch our lives fall apart again, so I started wandering. A couple miles from home, I discovered an untouched tract of desert divided by a band of old trees. In the heart of these trees was Witches' Castle, which was what we called the ruins of some long-lost rancher's house. I usually went there by myself, but on this day I went with my friend Stott. We had been planning all week long to hunt the heavy cottonwoods that hid birds whose calls we would hear if we arrived early or late in the day.

Stott's history is a mystery to me: I knew him for such a short time and at an age when identity is a glimmer, the gleam of a knife-blade, of teeth. He had a simian beauty. When I think of him now, I see a lean-limbed, hairless monkey wearing braces, Homo erectus with a vast knowledge of swearwords, and almost every word spoken in a burst of laughter. In class the girls brushed his hair, which was thin and fine as silk. However beautiful his hair may have been, when he had a gun in his hands, Stott was deadly. This fact was understood the way many unspoken things are understood between children.

Before the hunt, Stott and I checked and loaded the guns

while we smoked his mother's cigarettes. Stott had sweat in the dish of his throat, as though someone had placed a gold coin there. We sat on what we thought was the basement stairwell of the Witches' Castle. Mud and rock had flooded it, and we believed that was all that separated us from a luxurious, dimly lit room deep underground. I spent an afternoon digging it out with a kid named Wilken. Blue eyes. Perfect teeth. Lots of guns. Dead. He was killed in a motorcycle accident. Another useful fact: when the human head and a granite boulder collide, the boulder remains structurally unchanged. Something about young people dying makes me cynical. This reaction is no less remarkable than the relative non-reaction I have to the idea of older people dying. They had their chance and lived long lives. In terms of what we privilege and what we don't, biology is the great leveler. If Wilken weren't dead, maybe he'd be playing Ping-Pong or watching his wife sleep through a dream. He sat behind me in gym, Indian-style. I don't think there was one Native American in the whole school. Wilken and I killed a magpie together. We counted to three and shot. That night I was so excited I couldn't sleep.

This was before I knew Stott. The cigarettes made him sick. I poured the BBs down the throat of the gun and marveled at the size of his mouth. "How many rows of teeth do you have in there?" I told Stott to lean against the tree and breathe deep. "Put out that god-damn cigarette," he said, turning away. I didn't tell him he was green as pond water, or that he looked pretty when he grinned. When Stott spoke, it was as if he were trying to change water into words, which plunged from his lips like upside-down needles. A few minutes later, the sandy color returned to his face. Stott said he saw something running through the grass. He went his way and I went mine.

I wonder if Stott remembers how I saved his life. Or if not his life, his feet. Or if not his feet, his toes. One toe. Maybe all I saved him from was a little fear on a sunny day in wintertime. We had wandered deep into the fields behind my house.

Stott's boots had iced up and he was whining for home. I sat him in the sun and picked the dead grass that poked through the snow like the frozen tails of mice. I made a small fire and helped him take off his boots. He cried and cried like they do in church until I told him it was bad luck to cry near the fire. "Why?" he asked, sobbing and rubbing his feet. "Because tears are water," I said, matter-of-factly. I had never heard of such a thing, but I knew how superstition worked. Like a motor starting up, Stott began to cry again. I couldn't put the fear of superstition in him, so I gave him some cheese and crackers and built up the fire.

That day at the castle, Stott didn't need any help from me. A few minutes after he had disappeared into the trees, I heard him laughing and shouting "holy fuck," and this-that-and-the-other-thing. I hadn't made a kill all day so I was peeved when he walked out of the trees carrying, of all things, a male ring-necked pheasant. Stott could have carried out an emperor penguin and I wouldn't have been more surprised. I had been hunting for years and hadn't come across anything like this pheasant. It reminded me of the gold-egg-laying hen in "Jack and the Beanstalk." Once I saw the bird in the grass, still warm and the eyes not alive but not quite dead, either, I felt awe, jealousy, and mild sadness. "Head shot," Stott said, pushing aside the tiny feathers that had closed around the small black hole in the bird's skull. I tried to spread one of the bird's wings, but it wouldn't come easily so I let it go and it returned to the bird's side. "Well I better get him home," Stott said.

I cut south through the derelict apricot orchard that paralleled the castle, kicking up grasshoppers that flew into the grass and then crawled up the blades in preparation for another hasty departure. I sighted in a hopper that was about fifteen feet away and fired. I heard a small *tick* and the hopper was gone. I rested my gun on my shoulder and went to check on my handiwork. The hopper was lying in the grass like a half-closed pocketknife. I pinched his wings between my thumb and finger and looked him over. The BB had cut him in half just below the thorax. Even in that state the hopper spewed tobacco juice, his chelicerae working fiercely to pay out the foul-smelling fluid in a final effort to repel his attacker. I was unsettled by the hopper's churning mouthparts, so I dropped him in the grass and worked my way out of the orchard. I had almost cleared the trees when I heard settling wings.

Ten steps into the orchard I found the wing bearer: a robin had built her nest in one of the healthier apricot trees. The tree still had leaves, some of which had begun to darken into a hue of orange that made it difficult to spot the orange-breasted robin sitting quietly among them. She must have had a lot of confidence in this site because she had constructed her nest a mere five feet off the ground, well within reach of a curious biped like me. Or it may have been that she simply made do with what she had. I moved in for a closer look. I thought for sure she would take flight. I wanted her to, but she sat tight and didn't as much as blink an eye. I wish I could say I had no idea why she wouldn't leave her nest, or that I let her be, content with the knowledge that I could have killed her. Instead, I aimed at her heart and fired. I wasn't more than five feet away, so I heard the small chirp she made the moment she was shot and saw the tiny orb of blood erupt from her chest and spill out into the nest.

Even now, a quarter-century later, I want to run from this memory. I want to live without it. But what if I admit this memory belongs to the only life I will ever have and know? How does this possibility affect which moments I treat as precious? What if all pleasures and pains do not leave the Earth? What if they do not translate into the currency of some ineffable region beyond the edge of the mind? Then, as now, it seems I had very little choice but to lean my gun against the tree, reach up and pull the nest from the branches. I think I already knew that the robin was looking out for more than herself. But I wasn't prepared to find three just-hatched chicks, eyes still closed, silent mouths agape, their naked bodies wet with their mother's blood.

Even worse, I saw that one of the chicks was also dead. I figured that instead of directly hitting the robin, the BB first passed through the nest, on through the chick, and into the robin. I saw an old man walking toward me through the field. He had short white hair and wore a green, one-piece work suit with a zipper up the front. I could have easily outrun the old guy, but I stayed right there, kneeling in the grass, weeping. I had my back to him, but I could hear him getting closer and closer, until finally he stood beside me. I could not bring myself to look up at him because the sun was shining and I felt sick with fear and shame. So when, in a voice that trembled with age and not with anger, he asked me if I knew that what I had done was wrong, I nodded *yes* at his brown work boots. He stood there for a couple of minutes while I sobbed, perhaps because he wanted me to live with what I had done for as long as I could. I guess a couple minutes were enough because then he told me to get on home and not to come back. And so I didn't. Until now.

The other day I told my mother about this experience and she said it was evil. Although she immediately retracted her knee-jerk comment, given her immersion in a belief system that exchanges secular currencies for transcendental payouts, I understand her reaction. But I also regret it. For if, to borrow from Whitman, I had listened to what the talkers were talking, I may have never returned to this time in my life, nor indeed to any event that inspires shame or discomfort. I would not have had the courage to open the doors to the darkened rooms, nor light the flame to illuminate their contents.

When the talkers insist I honor parts of my life and deny the rest, I recall one summer evening when Nole and I walked to the lake near his cabin in the woods above Heber, Utah. Each fall, the spawning trout made their way up the small stream that fed the lake. Normally we would cast from shore, but that night we did not have tackle and we had a group of friends to feed, including Kim and Nole's girlfriend at the time, so we knelt in the tall grass that fringed the stream and fished for the fattened trout by hand.

I remember pushing up the sleeves of my wool sweater, the sweet and vaguely oily smell of the damp and beautifully woven hair as I crouched and watched the bubbling pool for the backs of trout. Nole was a few yards up and across the stream from me. He yelled something I could not quite hear and then plunged his hands into the dark autumn water. In one fluid motion he scooped a three-pound cutthroat out of the stream and into the grass behind him. Then he rose from his knees, walked into the grass, looked down at the trout and then at me as he held his hands twenty inches apart. The bangs of his white-blond hair leaked water down his face and lay in strands across his forehead. In the failing light, he looked severe, as though he had just been born and had always been there. I caught glimpses of the big trout flopping in the grass. A moment later, Nole eased the fish into his open hands, turned it upside-down, and struck its head on a nearby rock.

Where I knelt, the stream flowed over a small waterfall that kicked up a fine spray. My face was numb with cold so I had trouble mouthing the words "One more?" to Nole as he sat streamside and chewed a blade of frosty grass. He nodded and I sank my arms into the bubbling water and, hands open, waited. A trout bumped the outside of my hand. I felt the tail of another sweep my wrist. Then a trout swam into my hand the way a key fits into a lock. I could feel the slight pressure of its gill plates opening and closing against my thumb and fingers. I clamped down on the trout and lifted it out of the water. Although I could not hear him, I saw Nole laughing and pointing to his forearm. A strand of deep orange, mucus-covered roe had slipped out of the trout and down my arm. Nole brought his hand to his mouth and mimicked eating, but instead I scooped the delicate eggs into my hand and washed them into the devouring water.

We walked the dirt road back to the cabin. I had on light shoes and could feel the small stones. To the west was total dark, but a big moon was going to rise that night and the sky to the east was light enough to glimpse careening bats as they hunted for mayflies. Nole had threaded a piece of string through the gills of each trout and had given them to me and I enjoyed their heft. Back at the cabin I made a fire in the steel-rimmed pit just off the porch while Nole cleaned the trout. I held my hands over the young flame and thought how good it was to have fire and shelter and a warm, dear friend on a cold night. Nole dressed the fish in the light of a lamp on the porch table. I could see the roe spill from the trout's underside as Nole slid the knife down its length. He sliced through the gill cavity and then lifted the viscera and roe, flinging it into the woods. "Why should we be the only ones to eat well tonight?" he laughed.

Nole placed the fish together in tin foil with butter, salt, pepper, and wild onion. While they cooked I took off my shoes and warmed my feet by the fire. I remember the sky was clear and it was very cold, so the bunch of us sat close and held out our hands to the flames. The aspen leaves had fallen and the ground was slick and fragrant with them. I stepped away to take a leak and I could hear the mallards panic down on the pond. I wondered what hunted them as smoke drifted and spiraled into the woods. I could not see what lay beyond the light of our fire, but I had walked right into it and it was there with us, like desire and flesh and hunger. As I walked back to my circle of friends laughing and telling stories around the fire, I realized how hard it was going to be to leave all this life, its sweet, fleeting, and bloody richness. The most precious thing we can bear.

MAXIMILIAN WERNER

Maximilian Werner is an award-winning author of three books: *Black River Dreams*, a collection of literary fly-fishing essays; the novel *Crooked Creek*; and the memoir *Gravity Hill.* Mr. Werner's poems, fiction, creative nonfiction, and essays have appeared in several journals and magazines, including *Matter Journal: Edward Abbey Edition, Bright Lights Film Journal, The North American Review, ISLE, Weber Studies, Fly Rod and Reel*, and *Columbia*. He lives in Salt Lake City and teaches writing at the University of Utah.

ABOUT TORREY HOUSE PRESS

The economy is a wholly owned subsidiary
of the environment, not the other way around.
—Senator Gaylord Nelson, founder of Earth Day

Headquartered in Salt Lake City and Torrey, Utah, Torrey House Press is an independent book publisher of literary fiction and creative nonfiction about the environment, people, cultures, and resource management issues relating to America's wild places. Torrey House Press endeavors to increase awareness of and appreciation for the importance of natural landscape through the power of pen and story.

2% for the West is a trademark of Torrey House Press designating that two percent of Torrey House Press sales are donated to a select group of not-for-profit environmental organizations in the West and used to create a scholarship available to upcoming writers at colleges throughout the West.

Torrey House Press
www.torreyhouse.com

See torreyhouse.com for our thought-provoking *Evolved* Discussion Guide.

ALSO AVAILABLE FROM TORREY HOUSE PRESS

***Crooked Creek* by Maximilian Werner**

Sara and Preston, along with Sara's little brother Jasper, must flee Arizona when Sara's family runs afoul of American Indian artifact hunters. Sara, Preston, and Jasper ride into the Heber Valley of Utah seeking shelter and support from Sara's uncle, but they soon learn that life in the valley is not as it appears and that they cannot escape the burden of memory or the crimes of the past. Evoking the lyricism of Cormac McCarthy and the elegance of Wallace Stegner, Werner combines vivid characters with brilliant tension in this potent allegory set in the nineteenth century West.

***Tributary* by Barbara K. Richardson**

Willa Cather and Sandra Dallas resonate in Richardson's fearless portrait of 1870s Mormon Utah. This smart and lively novel tracks the extraordinary life of one woman who dares resist communal salvation in order to find her own. Clair Martin's dauntless search for self leads her from the domination of Mormon polygamy to the chaos of Reconstruction Dixie and back to Utah where she learns from Shoshone Indian ways how to take her place, at last, in the land she loves.

***Recapture* by Erica Olsen**

The stories in *Recapture* take us to an American West that is both strange and familiar. The Grand Canyon can only be visited in replica form. An archivist preserves a rare map of a vanished Lake Tahoe. A Utah cliff dwelling survives as an aging roadside attraction in California. By turns lyrical, deadpan, and surreal, Erica Olsen's stories bring us the natural world and the world we make, the artifacts we keep and the memories and desires that shape our lives.

***The Ordinary Truth* by Jana Richman**

When Nell Jorgensen buried her husband after a hunting accident in 1975, she buried a piece of herself, her relationship with her daughter, and more than one secret along with him. Now, thirty-six years later, her granddaughter Cassie intends to unearth those secrets and repair those relationships, but she's unprepared for what she finds. Set in the sparse and beautiful landscape of Nevada's Spring Valley and Schell Creek Mountains, award-winning author Jana Richman brings us an emotional journey of love, loss, and family steeped in the realities of the colliding urban and rural worlds of the West.

***Grind* by Mark Maynard**

Gritty and irreverent, eight linked stories set near Reno, Nevada, bring a colorful band of characters to vivid life. A truck driver with the I.Q. of a child drives his empty rig up and down the Interstate as an homage to his dead mother ("Deadheading"), two convicts seek redemption by helping a group of horses get their freedom beyond the gates ("Penned"), a schizophrenic homeless man hits it big on a belligerent slot machine and then disappears into the wilds along the Truckee River leaving behind the payout that would change his life forever ("Jackpot"). These stories play out within sight of the Mother Lode hotel and casino, a glass monolith that looms over downtown Reno and whose peeling façade represents the city's dwindling past and the uncertain future.

***The Scholar of Moab* by Steven L. Peck**

A mysterious redactor finds the journals of Hyrum Thayne, a high-school dropout and wannabe scholar, who manages to wreak havoc among townspeople who are convinced he can save them from a band of mythic Book of Mormon thugs and Communists. Though he never admits it, the married Hyrum charms a sensitive poet claiming that aliens abducted her baby (is it Hyrum's?) and philosophizes with Oxford-trained conjoined twins who appear to us as a two-headed cowboy. Peck's hilarious novel considers questions of consciousness and contingency, and the very way humans structure meaning.

***The Plume Hunter* by Renée Thompson**

When Fin McFaddin takes to plume hunting—killing birds to collect feathers for women's hats—to support his widowed mother, he finds danger, controversy, and heartache amid the marshes of nineteenth century Oregon. In 1885, hunters like Fin killed more than five million birds in the United States for the millinery industry, prompting the formation of the Audubon Society. The novel brings to life an era of American natural history seldom examined in fiction as it explores Fin's relationships with his lifelong friends as they struggle to adapt to society's changing mores.